T0199759

Introduction to
SYSTEMS
ECOLOGY

Applied Ecology
and Environmental Management

A SERIES

Series Editor
Sven E. Jørgensen

Copenhagen University, Denmark

Introduction to Systems Ecology
Sven E. Jørgensen

Handbook of Ecological Indicators for Assessment of
Ecosystem Health, Second Edition
Sven E. Jørgensen, Fu-Liu Xu, and Robert Costanza

Surface Modeling: High Accuracy and High Speed Methods
Tian-Xiang Yue

Handbook of Ecological Models Used in Ecosystem and
Environmental Management
Sven E. Jørgensen

ADDITIONAL VOLUMES IN PREPARATION

Introduction to
SYSTEMS
ECOLOGY

Sven Erik Jørgensen

CRC Press
Taylor & Francis Group
Boca Raton London New York

CRC Press is an imprint of the
Taylor & Francis Group, an **informa** business

CRC Press
Taylor & Francis Group
6000 Broken Sound Parkway NW, Suite 300
Boca Raton, FL 33487-2742

First issued in paperback 2019

ISBN-13: 978-1-4398-5501-0 (hbk)
ISBN-13: 978-0-367-86609-9 (pbk)

Library of Congress Cataloging-in-Publication Data

Jørgensen, Sven Erik, 1934-
  Introduction to systems ecology / Sven Erik Jorgensen.
     p. cm.
  Includes bibliographical references and index.
  ISBN 978-1-4398-5501-0
  1. Biotic communities. 2. Bioenergetics. 3. Thermodynamics. 4. Ecology--Philosophy. I. Title.

QH541.J668 2012
577.8'2--dc23
                                                                    2011042870

Visit the Taylor & Francis Web site at
http://www.taylorandfrancis.com

and the CRC Press Web site at
http://www.crcpress.com

Contents

PART 2   Properties of Ecosystems

Preface

This is to my knowledge the first book in systems ecology with the features of a textbook: illustrations, exercises, summaries, and emphasis on the most important statements. The reason is probably that there are several useful theoretical angles to systems ecology or ecosystem theories: the thermodynamics, the hierarchical organization, the network theory, and biochemistry or ecological stoichiometry, and it is difficult to integrate these theories into one holistic theory. I hope that it has been possible for me with *Fundamentals of Systems Ecology* to make a useful first attempt, but I would like to encourage the readers to give me feedback, because it would make it possible to make a better and improved second edition in a few years. Please, send me an e-mail if you have good ideas about what is missing, what could be presented and organized more clearly, and how the book generally could be improved. My e-mail address is msijapan@hotmail.com.

Systems ecology is used more and more in a number of ecological disciplines. We use it in ecological modeling, because you cannot model a system if you do not know the properties and the reactions of the system. We use it in ecological engineering, because you cannot engineer a system unless you know the system. We use it to choose ecological indicators, because selection of the best ecological indicators for the health assessment of an ecosystem requires, of course, that we know the system and its properties and possible shortcomings. We use systems ecology generally in ecological and environmental management, because ecosystem management requires that we know the reactions of the ecosystems to changed impacts. Therefore, there is an obvious need for a textbook in systems ecology that can be used as a general frame of reference in ecological disciplines.

Systems ecology has four main aspects: thermodynamics, hierarchical organization, network theory, and biochemistry. The textbook is built on all four combined with a holistic systems approach that attempts to understand the system and not go into the details that are integrated in the system. The last chapter of the book is devoted to the applications of systems ecology in the above-mentioned ecological disciplines, and it is my hope that it will inspire more ecologists to apply systems ecology more widely in the future. Only by application of the ecosystem theory in many different contexts will it be possible to improve systems ecology in the future. Two of my scientific friends have expressed their interest to translate the textbook into Chinese and Russian. It would be very beneficial for systems ecology, because it would allow many more researchers and managers to contribute to the progress of systems ecology and to the general application of ecology to understand nature and to make a better environmental management.

Sven Erik Jørgensen
Copenhagen
August 2011

1

Systems Ecology: An Ecological Discipline

The big bang was the start, 13.6 billion years ago.
The most characteristic constants for the created universe pointed toward
the evolution of life (see, for instance, Laszlo, 2003)

The ecological discipline, system ecology or ecosystem theory, is presented. An ecosystem theory enhances environmental management due to the possibilities to predict the reactions of ecosystems to changed conditions. The outline and the content of the book are presented. Emphasis on holistic approaches is maintained throughout the book.

1.1 What Is Systems Ecology?

Systems ecology focuses on the properties of ecosystems and tries to reveal them by the use of a systems approach—to see the forest through the trees. It could also be denoted ecosystem theory, because this ecological subdiscipline attempts to develop a theory that can be applied to explain the characteristic processes and reactions of ecosystems similar to how physics is able to explain physical phenomena and make quantitative predictions about how physical systems will react to well-defined perturbations. On the basis of about 25 basic physical laws, we are able to deduce a number of other laws and rules in physics, explain our observations, and calculate at least approximately the influence of well-defined factors on physical systems. It makes theoretical physics very useful, because it implies that we can predict what is happening quantitatively by introduction of well-defined changes without making observations or carrying out experiments. Figure 1.1 shows these aspects of theoretical physics.

If we similarly could develop an ecosystem theory, it would be possible at least approximately to predict how ecosystems would react to specific pollutants and well-defined changes, for instance, climatic changes. Ecological models have been developed to be

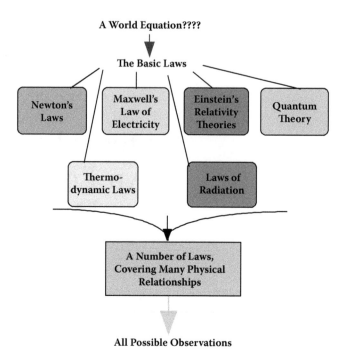

FIGURE 1.1 (See color insert.) In physics, it is attempted to explain all possible observations by use of about 25 fundamentals laws and a number of other laws and rules that are deduced from the 25 basic laws. It is a dream in physics to find a world equation that can be used to explain the approximate 25 basic laws, but the dream has not yet been realized.

able to describe the reaction in the modeled ecosystem of changing the external factors, denoted forcing functions, influencing the ecosystem. The reactions in the ecosystem are described by the use of state variables, which are the physical, chemical, and biological components of the ecosystem. The model needs to include what we call the ecological network, which is presenting the linkages between the components. The external factors, the forcing functions, or the impacts on an ecosystem may influence only one component at first, but due to the linkages, the influence will propagate throughout the entire ecosystem, and thereby influence directly or indirectly all the components. When we think in terms of models, it is possible to describe the benefits of having an ecosystem theory differently. The theory will be able to give us the changes of the state variables, when we change the forcing functions. This idea behind the use of models is illustrated in Figure 1.2. Notice that the forcing functions may be man-controlled functions, which means we are able to control them, or natural forcing functions, which means that we cannot control them, for instance, the climatic factors.

A model can be considered a synthesis of all that we know about the ecosystem and its forcing functions, including theoretical knowledge and observations. The more theoretical knowledge we have, the fewer observations are needed for building a good and well-working ecological model.

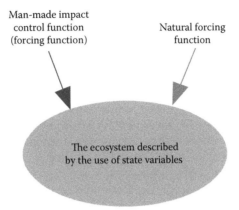

Man-made impact
control function
(forcing function)

Natural forcing
function

The ecosystem described
by the use of state variables

The model gives the relationship
between forcing functions and
state variables.

FIGURE 1.2 (See color insert.) Ecological models give the relationships between forcing functions and state variables.

An ecological model is built on all the knowledge that we have about the ecosystem and its forcing functions in the form of theoretical knowledge or in the form of corresponding observations of state variables and forcing functions. The better our theory is, the fewer observations are needed. In physics the theory is often so well developed that it is, for instance, possible to calculate the orbit in all details when we launch a rocket for Mars. We can predict where the rocket will land within 50 m, although the orbit of the rocket may be more than 60 million km—a realistic and possible figure for the distance to Mars. Our ecosystem theory is getting stronger and stronger, and we could, for instance, develop an ecological model that would be able tell us, without observations, the approximate concentration of mercury in the top carnivorous fish if we discharge a certain amount of mercury into an aquatic ecosystem with a known flow pattern. It would, however, need a much stronger ecosystem theory if we, without observations, should predict all relevant details how, for example, Lake Superior will change due to a change of the climate by an increase of the carbon dioxide concentration in the atmosphere from its present level of about 390 ppm to 500 ppm.

There is therefore a need for a stronger and more widely applicable ecosystem theory. In this textbook, a useful up-to-date ecosystem theory will be presented, and it will be demonstrated by many examples and illustrations how the theory could be applied to explain our observations and make quantitative calculations and predictions. There will, however, also be given examples that show that our present theory is not sufficient to explain all possible observations yet. There is a lot of room for further development of the ecosystem theory of today. On the other hand, it is very crucial to challenge the present theory wherever it is possible, because a steady use of the theory is the best method to define its shortcomings and see where further research is needed to reinforce the theoretical fundament.

1.2 The Holistic Approach

"The whole is greater than the sum of its parts" has been the core statement in system science and its many applications in all scientific disciplines. The statement was introduced by Aristotle and has been used by numerous scientists since. There are many examples that illustrate this basic system idea. Let us just mention a few to illustrate that a system consists of cooperative components that result in new and emerging properties. A Rembrandt painting is much more worthy than the sum of the canvas and the paints that were used to make the painting. The value of a human being when you account the value of the 25 elements that we consist of is probably not more than \$25, but because our 10^{14} cells work together as different organs that can coordinate and control many self-organizing biochemical processes, we can move, think, have feelings, be creative, understand, and so on. Similarly, an ecosystem is a cooperative unit of the organisms that make up the ecosystem. They are linked in a synergistic, cooperative, ecological network that can explain the adaptation, the development and evolution, the self-organization, and the resistance and flexibility to meet external disturbances, and we could even here include the beauty of the ecosystem. The importance of a holistic approach in system ecology is discussed in Chapter 14.

All systems with emerging properties cannot be described by listing the components and their properties, but it is necessary to capture, understand, and describe the emerging system properties. We have to see the forest through the trees. If we can, however, understand how ecosystems are working as systems, we will be able to develop an ecosystem theory that can be used to predict the consequences for an ecosystem of well-defined changes of forcing functions. The theory may be not strong enough to be able to make predictions with an acceptable low standard deviation completely without observations in all cases, but a good theory will make it feasible to give predictions that would require fewer observations to have an acceptable certainty. We can generally today understand and explain the reactions of ecosystems much better than we could 30–40 years ago due to the enhanced research in ecology and applied ecology, including the ecological subdisciplines, such as ecological modeling, ecological engineering, ecological indicators, and ecological informatics. We still need to supplement with observations in some cases, but the needs for observations are clearly lower than they were 30–40 years ago. System ecology is not at all as advanced today as theoretical physics, but a further development of this ecological field would be extremely beneficial for environmental management. A wider use of the theory with or without observations to supplement the theoretical considerations would inevitably give us an experience that would accelerate the development of system ecology.

1.3 Outline of the Book

The most basic and most widely applicable laws in science are probably the first, second, and third laws of thermodynamics. Ecosystems and all other systems and components have to follow these three laws. The three basic thermodynamic laws may be considered as constraints or restrictions on the possible ecological processes, which implies that the ecosystems have to obey these laws, and they will inevitably limit and determine the possible reactions and processes. It would facilitate our possibilities to make theoretical predictions on the reactions and processes of ecosystems by a combination of given forcing functions or impacts on a focal ecosystem, to include the constraints by the three thermodynamic laws in our considerations.

The three thermodynamic laws will be presented in Chapters 2–4, together with their immediate implications for ecosystems. The basic thermodynamic concepts are the basis for an understanding of these chapters and a general application of thermodynamics on ecosystems. The basic thermodynamic laws applied on ecosystems are often considered difficult to understand theoretically, but they are so fundamental that it is important to try to understand what they can tell us about ecosystems and their reactions. It has therefore been decided in this textbook about ecosystem theory to devote three chapters to present in sufficient detail the three basic thermodynamic laws and their ecological implications.

Chapter 5 presents some features of the basic biochemistry that all biological components have to follow. All biological components have a surprisingly similar biochemistry, which means that all organisms have approximately the same need for about 20–25 elements. Six of these elements are compulsory for living organisms. Furthermore, the reaction rates of biochemical processes set other conditions for life. So, the biochemistry defines additional constraints and restrictions on ecosystems.

Ecosystems have, however, been able to grow and develop in spite of energy and matter needing to be conserved, in spite of all processes being irreversible, and in spite of the basic biochemical composition of all biological components having to be obeyed. Ecosystems have been able to evade the thermodynamic and biochemical constraints by being open to energy flows, and by use of three well-defined possibilities to grow and develop: increase the biomass, increase the ecological network, and increase the information. Ecosystems can utilize their openness to get sufficient work energy and can furthermore select the pathways that move the ecosystem farthest away from thermodynamics, corresponding to the highest possible work capacity (also denoted eco-exergy) at the prevailing conditions, to achieve the most growth and development under the prevailing and constrained conditions. This last principle—to select the pathways that ensure the most exergy (work capacity)—can be considered a translation of Darwin's theory into thermodynamic terms, and it is presented in Chapter 7. The content of Chapters 6 and 7 can be formulated as a law, which is denoted the ecological law of thermodynamics (ELT), or sometimes also the fourth law of thermodynamics. Part 1 of the book covers what has been denoted the constraints on ecosystems, which are, on the one side, giving the conditions, namely, to obey the three laws of thermodynamics and the biochemically determined compositions and, on the other side, that the ecosystem must steadily grow and develop to ensure survival. The result of the constraints and the ability of the ecosystem to survive, grow, and develop in spite of the constraints, as expressed in ELT, has been that ecosystems have gained seven very basic, indispensable, and characteristic properties:

1. They are *open systems* far from thermodynamic equilibrium (this is an immediate consequence of the second law of thermodynamics).
2. They are *organized hierarchically*.
3. They have a *high diversity*.

4. They have *high buffer capacities*, which makes it difficult for forcing functions to change them radically.
5. The components are *organized in networks*, which allow recirculation and feedback regulation and ensure a high work energy (exergy) efficiency.
6. They have a *high information content*, embodied in the genomes of the organisms, which explains the very developed feedback and regulation mechanisms.
7. They have *emerging system properties* due to their very well-developed organization and structure far from thermodynamic equilibrium.

These seven basic properties are presented in Chapters 8–14, although the first property, openness, already is touched on together with the second law of thermodynamics, because it is very closely associated to the consequences of the second law of thermodynamics. The seven properties are a result of evolution. They are a result of the endeavor the ecosystems have made to survive, grow, and develop in spite of the conditions and the limitation of the three thermodynamic laws and the biochemically determined life processes.

The presented holistic systems ecology has a wide application by explaining ecological observations, in conservation biology, in ecological modeling, by assessment of ecosystem health and sustainability and in ecotechnology. Several illustrative examples of these applications are presented in Chapter 15 to demonstrate the importance of the application of an ecosystem theory by the subdisciplines of applied ecology. They are the fundament in environmental management and are toolboxes that provide environmental managers with possibilities to reduce the human impact on nature and its many different types of ecosystems. The tool boxes form a bridge between environmental management and ecology and make it possible to include ecological and ecosystem considerations in the control of pollution by environmental management plans.

Part 2 of the book covers Chapters 8–15, which discuss the seven basic properties and the application of the ecosystem theory in several ecological subdisciplines.

The ecosystem theory can be summarized as 14 propositions, which are presented in Chapters 2–14. The ecosystem theory is complete in the sense that it is probably able to cover all observations of ecological processes, but it needs, of course, to be challenged by wider applications that would reveal its weaknesses and thereby inspire improvements and produce a more solid theory about ecosystems and their reactions and processes.

The 14 propositions about the properties and reactions of ecosystems are:

1. Ecosystems, as all other systems, conserve matter and energy.
2. Ecosystems recycle all matter and partly all energy.
3. All processes in ecosystems are irreversible, produce entropy, and consume free energy (exergy or energy that can do work).
4. All the living components in ecosystems have the same basic biochemistry.
5. Ecosystems are open systems and require an input of free energy (exergy or energy that can do work) to maintain their function.
6. If the ecosystems receive more free energy to maintain their function, the surplus free energy will be applied to move the system further away from thermodynamic equilibrium.

7. Ecosystems offer many possibilities to move away from thermodynamic equilibrium and they select the pathway that moves the ecosystem farthest from thermodynamic equilibrium.
8. Ecosystems apply three growth forms: (a) growth of biomass, (b) growth of the network, and (c) growth of information.
9. Ecosystems are hierarchically organized.
10. Ecosystems have a high diversity in all levels of the hierarchy.
11. Ecosystems have a high buffer capacity toward changes.
12. The components of ecosystems work together in a cooperative network.
13. Ecosystems contain an enormous amount of information.
14. Ecosystems have emergent system properties.

1

Sciences: Basic for Systems Ecology

2

Conservation of Energy and Matter

The sun is the ultimate energy source for all activities on earth.
Without the sun, everything on earth dies.
There is no free lunch.

2.1 The Conservation Laws

The important conservation laws and a number of core thermodynamic functions, with emphasis on free energy covering the work capacity, are presented. The relationship between free energy and the chemical equilibrium is in this context very important. The conservation laws are applied in ecology to understand Liebig's law of minimum, bioaccumulation, biomagnification, and recycling.

The first law of thermodynamics expresses that energy is conserved. Energy cannot be destroyed or created. The law is often expressed mathematically by the following equation:

$$\Delta U = \Delta Q + \Delta W \qquad (2.1)$$

where U is the energy and ΔU is the increase of energy, ΔQ is the amount of heat received from the environment, and ΔW is the amount of work received from the environment. The equation expresses just a bookkeeping of the energy considering the two forms of energy: heat and the energy that can do work. The gain in energy is just a matter of how much energy the system receives from the environment either as heat or as work.

The principle of conservation of energy, called the first law of thermodynamics, was initiated in 1778 by Rumford. He observed that a large quantity of heat appeared when a hole is bored in metal. Rumford assumed that the mechanical work was converted to heat by friction. He proposed that heat was a type of energy that is transformed at the expense of another form of energy, here mechanical energy. It was left to J.P. Joule in 1843 to develop a mathematical relationship between the quantity of heat developed and the mechanical energy dissipated.

Two German physicists, J.R. Mayer and H.L.F. Helmholtz, working separately, showed that when a gas expands the internal energy of the gas decreases in proportion to the amount of work performed. These observations led to the first law of thermodynamics: energy can be neither created nor destroyed.

If the concept of internal energy, dU, is introduced:

$$dU = dQ - dW \ (ML^2T^{-2}) \tag{2.2}$$

where dQ is thermal energy added to the system, dU is increase in internal energy of the system, and dW is the work done by the system on its environment. Work done by the environment on the system would of course contribute positively to the energy of the system, as expressed in Equation (2.1).

This important law is often seen formulated in several different versions. The difference between the two formulations in Equations (2.1) and (2.2) is not only that Δ is used in (2.1) and d in (2.2), but that the work in (2.2) is the work done by the system, while in (2.1) ΔW is the work done on the system. Both versions are used in the scientific literature.

The principle of energy conservation can also be expressed in mathematical terms as follows:

U is a state variable, which means that $\int_1^2 dU$ is independent of the pathway 1 to 2.

In ecology there is practically no formation of energy from matter, but by nuclear processes it is possible to transform matter to energy. The radioactive processes are, however, insignificant in ecology. Einstein has, with his famous equation $E = mc^2$, indicated how much energy, E, it is possible to gain by destruction of the mass m. c is the speed of light, about 300,000 km/s.

As nuclear processes are not considered in ecology, matter is like energy conserved, which means:

$$\text{Accumulation} = \text{input} - \text{output} \tag{2.3}$$

This means as for energy, it is also possible for matter to use a bookkeeping of the exchange of matter with the environment. The amount of matter gained minus the amount of matter lost to the environment will tell us directly how much matter is accumulated— added to the system. The equation is sometimes expressed by the use of concentrations:

$$dC/dt = (\text{input} - \text{output})/V \tag{2.4}$$

where C is the concentration in the system and V is the volume of the system. In this case, the concentration is expressed per unit of area, and V is replaced by the area, A.

If the law of mass conservation is used for chemical compounds that can be transformed to other chemical compounds, then Equation (2.4) must be changed to:

$$V*dc/dt = \text{input} - \text{output} + \text{formation} - \text{transformation} \ (MT^{-1}) \tag{2.5}$$

The principle of mass conservation is widely used in the class of ecological models called biogeochemical models. As we do not consider radioactive processes in ecology—or very rarely—the mass conservation principle can be used for all elements. The equation is set up for the relevant elements, e.g., for eutrophication models for C, P, N, and perhaps Si (see Jørgensen, 1976a, 1976b, 1982; Jørgensen et al., 1978).

To summarize: Ecosystems conserve matter and energy, which means that it is possible by bookkeeping of the exchange processes between the system and the environment to calculate the system's gain or loss of energy and matter and even for each element.

The first law of thermodynamics may also be stated in terms of the universal human experience, that it is impossible to construct a perpetual motion machine that can generate energy.

2.2 Other Thermodynamic Functions

The work can take various forms. If we presume that the work is carried out as an increase in volume, ΔV, under constant atmospheric pressure, p, for the system, we get:

$$\Delta U = \Delta Q - p \, \Delta V \qquad (2.6)$$

This equation is applied to define a new function, named enthalpy, or heat content, H:

$$H = U + pV \qquad (2.7)$$

$$\Delta H = \Delta U + \Delta(pV) \qquad (2.8)$$

Assuming that the ideal gas law is valid, we have that $\Delta H = \Delta U + \Delta(nRT)$.

We are often interested in the enthalpy change when the process occurs in the system at a fixed external pressure, for instance, 1 atm, as is the case with many biochemical processes. Under these circumstances the work done by the change of the volume (for instance, by generation of a gas) cannot be utilized, and therefore the change in enthalpy becomes as relevant as the change in energy of the considered system. Heat of a reaction has biochemical significance. If heat is produced, the process is named exothermic. Decomposition of the organic matter in food is an exothermic process that provides heterotroph organisms with the chemical energy that is needed for the maintenance of life. Endothermic processes require addition of heat. A typical endothermic process is the formation of ATP (adenosine triphosphate) from phosphate and ADP (adenosine diphosphate), expressed in biochemistry in the following equation:

$$ADP + P + 42 \text{ kJ/mole} = ATP \qquad (2.9)$$

The opposite process, formation of P and ADP from ATP, yields similarly 42 kJ/mole. ATP is applied in the organisms as a suitable unit of energy to be applied in the cells, wherever it is needed to carry out biochemical processes. ATP delivers the energy to a variety of processes: synthesis of cellular macromolecules, synthesis of membranes, cellular movements, electrical potential across membranes, transport of molecules against a concentration gradient, and maintenance of a suitable temperature, to mention a few ecologically important processes.

The heat of reaction may be by either constant pressure or constant volume, which implies that the heat of reaction becomes, respectively, the change in enthalpy and the change in internal energy. In reactions involving only liquids or solids, ΔV is in most cases negligible, and there is therefore no practical difference between the two heats of reaction. A convenient standard state for a substance may be taken to be the state in which it is stable at 25°C and at 1 atm pressure. The standard enthalpy of any compound is the heat of reaction by which it is formed from its elements, reactants, and products all being in the standard state. A superscript of 0 indicates the standard heat of formation with reactants and products at 1 atm pressure.

The classical thermodynamics has introduced two more energy-based functions: the work function, A, and free energy, G. They have been introduced because it is desirable to obtained criteria for thermodynamic equilibrium under practical conditions, which means that the temperature is approximately constant in addition to either constant volume in bomb calorimeters or constant pressure in chemostats and ecosystems. The two functions are defined according to the following equations:

$$A = U - TS \tag{2.10}$$

$$G = U + pV - TS \tag{2.11}$$

or by differentiating:

$$dG = d\acute{U} + pdV + Vdp - TdS - SdT \tag{2.12}$$

and since $dU = T\,dS - pdV$, we obtain, when no chemical reactions take place:

$$dG = V\,dp - S\,dT \tag{2.13}$$

For a change at constant temperature and volume:

$$\Delta A = \Delta U - T\Delta S \tag{2.14}$$

$$\Delta G = \Delta H - T\Delta S = \Delta U - p\Delta V - T\Delta S \tag{2.15}$$

where S is the function named entropy defined by $dS = dQ/T$. The introduction of entropy is often associated with the second law of thermodynamics, which is presented in the next chapter. Most chemical processes in laboratories or in organisms are carried out at constant temperature and pressure. This implies that $dG = -dW$. We are often presuming constant temperature and pressure for processes in ecosystems. At thermodynamic equilibrium, all gradients are eliminated according to the definition. This means that no work can be performed. This implies at constant temperature and pressure that $dG = 0$, which is an important consequence of the energy conservation law. A general dynamic equilibrium, but not at thermodynamic equilibrium, is possible by equal process rates in opposite directions to ensure that steady state is maintained.

Standard free energies of formation of chemical compounds, $\Delta G°$, are very important for calculation of the chemical equilibrium. The standard free energy of formation of a compound is the free energy of the reaction by which it is formed from its elements, when all the reactants and products are in the standard state (1 atm pressure and 25°C). For the standard state 20°C (or sometimes, as indicated above, 25°C) and 1 atm pressure can be used—conditions we often find in ecosystems. The standard free energy of elements is according to this definition 0. Free energy equations can be added and subtracted just as thermochemical equations. This implies that the free energy of any reaction can be calculated from the sum of the free energies of the products minus the sum of the free energies of the reactants:

$$\Delta G° = \Sigma\ G°\ (\text{products}) - \Sigma\ G°\ (\text{reactants}) \tag{2.16}$$

Chemical and physical handbooks contain long tables of $G°$. Free energy describes the chemical affinity under conditions of constant temperature and pressure: $\Delta G = G(\text{products}) - G(\text{reactants})$. When the free energy is 0 there is no net work obtainable by any change or reaction at constant temperature and pressure. Consequently, the system is in a state of equilibrium.

When the chemical energy change is positive for a proposed process, an input of free energy (energy that can do work) is needed to realize the reaction; otherwise it cannot take place. When the free energy change is negative, the reaction can proceed spontaneously by providing useful net work. The system loses free energy (work).

The levels of free energy for various biologically important nitrogen compounds are shown in Figure 2.1. Protein contains the most free energy, and when it is oxidized to nitrate, the free energy difference is released. The figure shows how the free energy stepwise is decreased from protein, to polypeptides, to amino acids, to ureas, to ammonium, to nitrite, and finally to nitrate, which is the end product for aerobic decomposition of proteins.

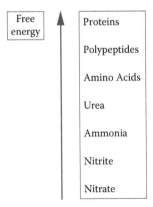

FIGURE 2.1   (See color insert.) The levels of free energy for various biologically important nitrogen compounds.

It can be shown for ideal gas reactions that the equilibrium constant K_p for the considered process is related to $-\Delta G^\circ$ by the following equation:

$$-\Delta G^\circ = RT \ln K_p$$

The expression for K_p can best be illustrated by an example. The process $H_2 + CO_2 = H_2O + CO$ has the equilibrium constant:

$$K_p = p\,(H_2O)\,p\,(CO)/p(H_2)\,p(CO_2) \tag{2.17}$$

However, very few chemical reactions follow the ideal gas laws. It has therefore been necessary to introduce a new variable named the chemical potential, μ_i, defined as $(\partial G/\partial n_i)T,p,n_j$. In other words, it is the change of the free energy with the change in the number of molecules of the ith component, n_i, when the temperature, the pressure, and the number of molecules of all other components are kept constant. Using the chemical potential, we obtain the following expression for dG (see Equation (2.13)):

$$dG = -S\,dT + V\,dp + \sum_i \mu_i dn_i \tag{2.18}$$

or at constant temperature and pressure:

$$dG = \sum_i \mu_i dn_i \tag{2.19}$$

At thermodynamic equilibrium $dG = 0$. It implies that for a gas with the partial pressure p_i

$$\mu_i - \mu_i^\circ = RT \ln p_i \tag{2.20}$$

It is convenient here to introduce a new function called fugacity, f, of considered substances. Fugacity is defined from the following equation:

$$\mu_i - \mu_i^\circ = RT \ln f_i/f_i^\circ \tag{2.21}$$

A standard state may, however, be defined as the state of unit fugacity, as the standard state for ideal gases was the state of unit pressure. It is now possible to set up an expression for the equilibrium constant, which is true in general not only for real (nonideal) gases, but also for substances in any state of aggregation. If we consider the process aA + bB = cC + dD, we get:

$$K_f = \{C\}^c\{D\}^d/\{A\}^a\{B\}^b \tag{2.22}$$

where {} indicates the fugacity. Equations (2.16) and (2.17) yield:

$$-\mu_i^\circ = RT \ln K_f \tag{2.23}$$

It can be shown that the fugacity can be replaced by concentrations (pressures for gases) in many calculations with a good approximation. For solutions it is possible to find the fugacity by multiplying the concentration with an activity coefficient that can be found by empirical equations. For aquatic solutions the fugacity coefficient is close to 1.00 by a total concentration of dissolved matter of less than 1 g/L. This means that fugacity has to be applied for marine ecosystems, because the salinity is usually more than 1 g/L.

It is emphasized in this context that these thermodynamic calculations of equilibrium constant and standard heat and free energy are also valid for biochemical processes and are the reactions of interest in an ecosystem.

We distinguish among different forms of energy. All forms can be described as a quantitative (extensive) variable times a qualitative (intensive) variable. The extensive and intensive variables for different energy forms are summarized in Table 2.1.

Work can be performed when the extensive variable is changed from one level of the intensive variable to another, and the work equals the extensive variable times the difference of the intensive variable between the two levels. As energy is conserved, work implies that one energy form is transferred to another energy form.

Energy of a considered system is often defined as the ability to do work. This definition is, however, in disagreement with the second law of thermodynamics, as we shall see in the next chapter. It is presumed in this context that the system is transferred to the level where the intensive variable is 0, whereby the work performed becomes equal to the energy content of the system. Heat energy is, however, different from the other energy forms in the sense that the complete utilization of the energy to do work is impossible. The heat energy can be utilized in the so-called Carnot cycle to do work by transfer from a warmer system to a colder system. The efficiency becomes $T - T_o/T$, which means a complete utilization of the energy to do work requires that $T_o = 0$, and 0 K is impossible to achieve for the cold reservoir without investment of more work than obtained by the Carnot process. The limitations of energy transfer by 0 K will be further discussed in the next chapter, when the third law of thermodynamics is presented.

TABLE 2.1 Different Forms of Energy and Their Intensive and Extensive Variables

Energy Form	Extensive Variable	Intensive Variable
Heat	Entropy (J/K)	Temperature (K)
Expansion	Volume (m^3)	Pressure (Pa = kg/s^2 m)
Chemical	Moles (M)	Chemical potential (J/moles)
Electrical	Charge (amp s)	Voltage (volt)
Potential	Mass (kg)	(Gravity) (height) (m^2/s^2)
Kinetic	Mass (kg)	0.5 (velocity)^2 (m^2/s^2)

Note: Potential and kinetic energies are denoted by mechanical energy.

TABLE 2.2   Average Elemental Composition
of Freshwater Plants, Wet Weight Basis

Element	Plant Content (%)
Oxygen	80.5
Hydrogen	9.7
Carbon	6.5
Silicon	1.3
Nitrogen	0.7
Calcium	0.4
Potassium	0.3
Phosphorus	0.08
Magnesium	0.07
Sulfur	0.06
Chlorine	0.06
Sodium	0.04
Iron	0.02
Boron	0.001
Manganese	0.0007
Zinc	0.0003
Copper	0.0001
Molybdenum	0.00005
Cobalt	0.000002

Source: Based on Wetzel, R.G., 1983. *Limnology*,
Saunders College Publishing, Philadelphia.

2.3 Liebig's Law of Minimum

The conservation laws have major environmental consequences. As we do not consider
nuclear processes, we can use the bookkeeping principle for all 90 naturally occurring
elements. This means that if an element needed for the growth of biomass is used up, it
cannot be created inside the ecosystem but must be added from the environment. About
20–25 elements are needed for most organisms, and the growth of an organism must
stop when the element that is present in the ecosystem in the smallest amount relative to
the needs for the considered organism is used up.

The needs are not necessarily a constant and fixed concentration. Table 2.2 gives the
average concentration for freshwater plants (wet basis). In the table it is indicated, for
instance, that the phosphorus concentration is 0.08% on a wet basis, which means that
if the dry matter content is 10%, the average phosphorus concentration is 0.8%. Many
plants have from about 0.4 to about 2.0% phosphorus on a dry basis. This implies that
when there is no more phosphorus and a plant has reached the lowest feasible percentage
of phosphorus, 0.4%, the growth must inexorably stop.

The environment rarely has exactly the chemical composition required for growth, which
means the element in least supply compared with the needs determines the limits to growth.

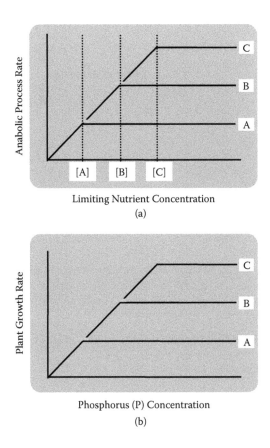

FIGURE 2.2 (a) General relationship between the rate of an anabolic process and the concentration of a limiting nutrient for that process. At limiting concentrations the process proceeds at the limited rates allowed. Three limiting concentrations ([A], [B], [C]) of a single nutrient are shown, and three correspondingly limited process rates (A, B, C). (b) Plant growth illustration of Liebig's law of the minimum. Phosphorus (P) concentration is plotted against growth rate. Under nonlimiting conditions growth is linearly related to P concentration. At concentrations where elements other than phosphorus become limiting, higher P will not increase growth. The three levels A, B, and C in this case correspond to three different growth-limiting concentrations of elements other than phosphorus.

This is Liebig's classic law of the minimum. If a nutrient concentration is greater than 0, that corresponds of course to no growth; the unconstrained relation between the process and concentration will be linear. If less, then the process will be bounded by the limiting level (Figure 2.2a).

Liebig's law is often used in ecological modeling to account for growth of plants, including phytoplankton. A constant stoichiometric approach is presumed. The most frequently applied equation for this approach is

$$\text{Growth} = \mu\text{max} \left(PS/(PS + k_p) \right) \tag{2.24}$$

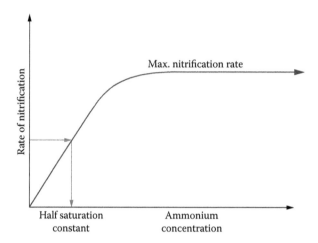

FIGURE 2.3   (See color insert.) Graph of the Michaelis-Menten equation. In this case, the rate of nitrification vs. the concentration of the substrate, ammonium, is shown. The same graph is obtained for the growth rate of phytoplankton vs. the inorganic reactive phosphorus concentration, provided that phosphorus is the limiting nutrient.

If two nutrients are limiting, the following equation may be applied:

$$\text{Growth} = \mu\text{max}^*\text{min } (NS/(k_n + NS), (PS/(PS + k_p) \tag{2.25}$$

where μmax is the maximum growth rate and k_n and k_p are Michaelis-Menten half saturation constants. PS and NS are the concentrations of dissolved reactive inorganic phosphorus and nitrogen, respectively. They presume that phosphorus and nitrogen (and maybe also silica and carbon) are taken up in a given stoichiometric ratio. A number of ecological processes are described by the Michaelis-Menten equation.

If multiple limiting factors are in play, further addition of only one of them will not influence growth (Figure 2.2b). Liebig's law is a direct consequence of the conservation of matter.

Figure 2.3 shows the graph of the Michaelis-Menten equation (2.24). The nitrification rate is plotted vs. the ammonium concentration. The microbiological oxidation of ammonium to nitrate follows. The chemical process is

$$NH_{4+} + 2O_2 \rightarrow NO_{3-} + H_2O + H^+ \tag{2.26}$$

Example 2.1

It is often discussed in lake management which nutrient is limiting the phytoplankton growth and thereby the eutrophication (too high primary production of phytoplankton due to a too high nutrient concentration). On average, the concentration of nitrogen in phytoplankton is seven times the concentration of phosphorus. For a considered lake, drainage water from agriculture contains 25 times as much nitrogen as phosphorus, while the wastewater discharged to the lake has only 4 times as much nitrogen as phosphorus, below which ratio of wastewater to drainage water, R, would phosphorus not be the limiting nutrient?

Also find R if the wastewater has only three times as much nitrogen as phosphorus.

Solution

Let us call the ratio of wastewater to drainage water where P is limiting RS. The following R would correspond to RS:

$$R4 + 25 = 7(R + 1) \text{ or } 3R = 18; R = RS = 6$$

If the ratio is less than 6, P becomes limiting, while if the ratio is more than 6, nitrogen is the limiting element. For three times, we have the following equation:

$$R3 + 25 = 7(R + 1) \text{ or } R = RS = 4.5$$

This means that if the ratio is more than 4.5, P becomes limiting, and if the ratio is less than 4.5, nitrogen is the limiting element.

Example 2.2

The transformation of solar energy to chemical energy by plants conforms with the first law of thermodynamics (see Figure 2.4).

Translate Figure 2.4 to energy balances.

Solution

Incoming sunlight = 1.97 GJ m^{-2} y^{-1} = solar energy assimilated + (solar energy reflected + solar energy applied for evaporation)

Solar energy reflected + solar energy applied for evaporation = 1.95 GJ m^{-2} y^{-1}

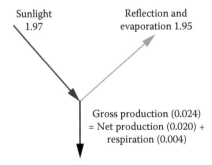

Sunlight 1.97

Reflection and evaporation 1.95

Gross production (0.024)
= Net production (0.020) +
respiration (0.004)

FIGURE 2.4 Fate of solar energy incident upon the perennial grass herb vegetation of an old field community in Michigan. All values in GJ m^{-2} y^{-1}.

Solar energy assimilated = 0.024 GJ m⁻² y⁻¹

Solar energy assimilated by plants = gross production =
chemical energy of plant tissue growth + heat energy of respiration

Chemical energy of plant tissue growth = net production = 0.020 GJ m⁻² y⁻¹

Respiration = the energy applied by the plants to maintain the
structure far from thermodynamic equilibrium = 0.004 GJ m⁻² y⁻¹

2.4 Bioaccumulation and Biomagnification

Figure 2.5 shows application of the mass balance principles on a heterotroph organism, which is dependent on food as the energy source. The food minus the nondigested part (feces) and the components taken up from water or air (for most organisms it is oxygen) are the inputs. The outputs are the natural mortality, the excretion, and the possible use of the organism as food for the next tropic level. The growth corresponds to accumulation = inputs – outputs. Figure 2.5 emphasizes that a part of the available food may not be utilized and that the amount of food utilized by the body is denoted the assimilated food, which is either used as an energy source by oxidation and respired or used for growth or excreted. The fraction of food assimilated varies from organism to organism and from food source to food source, but on average will be about two-thirds of the food assimilated.

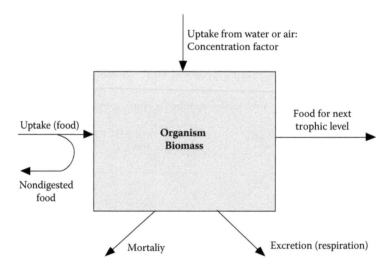

FIGURE 2.5 (See color insert.) The mass balance principle applied on an organism. The inputs are food minus the nondigested food, and the outputs are mortality, excretion, and possible use of the food for the next level in the food chain.

The mass flow through a food chain is mapped using the mass conservation principle. The food taken in by one level in the food chain is used in respiration, waste food, undigested food, excretion, and growth, including reproduction.

If the growth and reproduction are considered as the net production, then it can be stated:

$$\text{Net production} = \text{utilized food} - \text{respiration} - \text{excretion} - \text{feces} \qquad (2.27)$$

The ratio of the net production to the intake of food is named the net efficiency. The net efficiency is dependent on several factors, but is often as low as 10–20%.

Food contains chemical energy, and the shown mass balances can easily be converted to energy balances (see Figure 2.6):

$$F = A + UD = G + H + UD + EX, (ML^2T^{-2}) \qquad (2.28)$$

where F = the food intake converted to energy (Joule) = utilized food, A = the energy assimilated by the animals = digested food, UD = undigested food or the chemical energy of feces, G = chemical energy of animal growth, H = the heat energy of respiration (it is the energy the organism uses to maintain a structure far from thermodynamic equilibrium), and EX = excreted food.

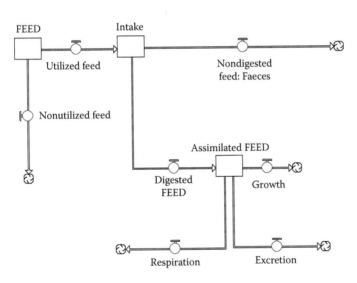

FIGURE 2.6 (See color insert.) The available food may be either utilized or not utilized. The food eaten (intake) is either digested or not digested. The assimilated food is used for growth, respiration (maintenance of the structure far from thermodynamic equilibrium), or excreted. The illustration has adopted the symbols used by the software STELLA, where boxes are state variables and pipelines with valves are processes.

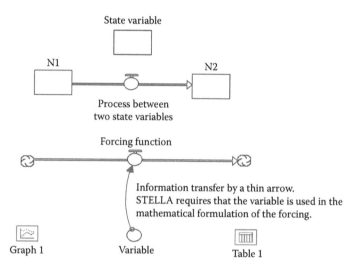

FIGURE 2.7 (See color insert.) The components used for construction of a STELLA diagram: state variables, forcing functions, processes, information transfer, and auxiliary variables. It is possible to get the results as tables and graphs, as indicated in the figure. The diagram is useful to illustrate the dynamics of ecosystems.

These considerations pursue the same lines as those mentioned in context with Equation (2.9) and Figures 2.5 and 2.6, where the mass conservation principle is applied. Biomass can be translated into energy, and this is also true for transformations through food chains. Ecological energy flows are of considerable environmental interest, as calculations of biological magnifications are based on energy flows. Throughout the book, we will apply a STELLA diagram to illustrate a conceptual diagram of an ecological model. The components applied for construction of the diagram are shown in Figure 2.7.

The assimilation of toxic substances may be very different from the assimilation of food in general, which is indicated at about 67%. Only about 5–10% of heavy metals are assimilated, while the assimilation of an organic toxic substance with a high K_{ow} (the ratio of the solubility in octanol to the solubility in water for the considered organic substance), such as, for instance, DDT and PCB, is 90% or more.

An organism takes up a toxic substance either by the food or from water or air. The ratio of the concentration of the toxic substance in the organism to the concentration in water or air is denoted the concentration factor. Toxic compounds in the food are unlikely to be lost through respiration and excretions because they are much less biodegradable than the normal components in the food. This being so, the net efficiency of toxic matter is often higher than for normal food components, and as a result, some chemicals, such as chlorinated hydrocarbons, including DDT and PCB, will be bioaccumulated—indicating that the concentration of the toxic substance is most likely higher in the organisms than in the food and water or in the air. If we consider the entire food chain, where one tropic level is a food source for the next trophic level, then the concentration of the toxic substance is magnified through the food chain. This phenomenon is

TABLE 2.3 Biological Magnification

Trophic Level	Concentration of DDT (mg/kg dry matter)	Magnification
Water	0.000003	1
Phytoplankton	0.0005	160
Zooplankton	0.04	~13,000
Small fish	0.5	~ 167,000
Large fish	2	~667,000
Fish-eating birds	25	~8,500,000

Source: Data after Woodwell, G.M., et al., *Science*, 156, 821–824, 1967.

TABLE 2.4 Concentration of DDT (mg per kg dry matter)

Atmosphere	0.000004
Cultivated soil	2.0
Freshwater	0.00001
Seawater	0.000001
Grass	0.05
Aquatic macrophytes	0.01
Phytoplankton	0.0003
Invertebrates on land	4.1
Invertebrates in sea	0.001
Freshwater fish	2.0
Sea fish	0.5
Eagles, falcons	10.0
Swallows	2.0
Herbivorous mammals	0.5
Carnivorous mammals	1.0
Human food, plants	0.02
Human food, meat	0.2
Man	6.0

called biomagnification and is illustrated for DDT in Table 2.3. DDT and other chlorinated hydrocarbons have an especially high biological magnification because they have
a very low biodegradability and are only excreted from the body very slowly, due to
dissolution in fatty tissue. These considerations also can explain why pesticide residues
observed in fish increase with the increasing weight of the fish. As humans are the last
link of the food chain, relatively high DDT concentrations have been observed in human
body fat (see Table 2.4).

If it is presumed that there is no mortality and the organism is not a food source for
the next trophic level, then we can set up the following equation for the concentration of
a toxic substance in the organism:

$$dTx/dt = \text{(daily) input} - kTx \qquad (2.29)$$

The daily input is an addition of the input of the toxic substance via the food and water or air. Tx is the amount of toxic substance in the organisms, for instance, expressed in g.

$$\text{At steady state: } dTx/dt = 0 \text{ or } kTx = \text{input or } Tx = \text{input}/k \qquad (2.30)$$

$$\text{Concentration can be found} = Tx/\text{biomass} \qquad (2.31)$$

where the biomass may be a function of time considering, for instance, growth of the organism.

If we consider the growth, the following equations are valid:

$$dW/dt = aW^b - rW^d \qquad (2.32)$$

In accordance to allometric principles, we have (see Peters, 1983):

$$b \approx 0.67; d \approx 0.75$$

The allometric principles will be treated in more detail in Chapter 8.

Uptake from water can be found as $CF^*C_w{}^*dW/dt$, where CF is the concentration factor for the toxic substance, and C_w is the concentration of the toxic substance in water or air. The expression presumes that CF is approximately a constant and that the toxic substance is taken up from the media corresponding to the growth to maintain the value of CF.

$$\text{Uptake from food} = aW^{b*}Cf^*\text{eff} \qquad (2.33)$$

where eff is the assimilation efficiency of the toxic substance, which is 0.05–0.1 for heavy metals and >0.9 for DDT and PCB. Excretion is presumed a first-order reaction, using the symbol exc for the excretion coefficient (1/24 h):

$$\text{Excretion} = \text{exc}^*Tx$$

Equation (2.29) can now be written as

$$dTx/dt = BCF^*C_w{}^*dW/dt + aW^{b*}C_f{}^*\text{eff} - \text{exc}^*Tx \qquad (2.34)$$

Corg = Tx/W = f (time), where Corg is the concentration of the toxic substance in the organism.

The equations calculate the bioaccumulation and they must be applied repeatedly to account for biomagnification.

Example 2.3

A food chain consists of four levels. For each of the four levels the fraction of food (dry matter basis) assimilated can be considered to be two-thirds of the food intake. The amounts of food intake for the four levels per unit of time are, respectively, 10, 2, 1, and 0.36, expressed in kg/24 h. The respiration is 20% of the biomass per 24 h. The food for the first trophic level contains 2 ppm of a toxic substance, which is assimilated 90% by all four trophic levels. The toxic substance is excreted by a rate of 0.01 per unit of time for all four trophic levels.

Calculate the concentrations of the toxic substance in the four levels at steady state and indicate the series of biomagnification.

Solution

At steady state inputs equal outputs for each trophic level. This means that the assimilated food equals food for the next trophic level—the respired biomass (which is equal to 0.2 × biomass). As we know the assimilated food and the food for the next trophic levels, we can find biomass as five times the difference between assimilated food and food for the next trophic level.

This means:

Biomass level 1 = 5(10*2/3 – 2) = 23.5
Biomass level 2 = 5(2*2/3 – 1) = 1.7
Biomass level 3 = 5(1*2/3 – 0.36) = 1.55
Biomass level 4 = 5*0.36*2/3 = 1.2

Similar equations can be developed for the toxic substances in each level (unit applied ppm × kg = mg).

Assimilated toxic substance – excreted toxic substance (which is equal to 0.01 times the toxic substance in the trophic level) – toxic substance in the food for the next trophic level (which is equal to the food times the concentration of the toxic substance in the trophic level). Let us indicate the amount of toxic substance in the trophic levels as X1, X2, X3, and X4. The concentration of toxic substance in the four levels is the toxic substance of the level divided by the biomass.

Toxic substance level 1 = 0.9(10*2) = 18 mg = 0.01X1 + 2X1/23.5 = 0.097X1 or X1 = 195 mg. The concentration of toxic substance in level 1 = 195/23.5 = 8.7 ppm (mg/kg).
Toxic substance level 2 = 0.9(2*8.7) = 15.7 mg = 0.01X2 + 1X2//1.7 = 0.6X2 or X2 = 26 mg. The concentration of toxic substance in level 2 = 26/1.7 = 15 ppm (mg/kg).
Toxic substance level 3 = 0.9(1*15) = 13.5 mg = 0.01X3 + 0.25X3/1.55 = 0.17X3 or X3 = 80 mg. The concentration of toxic substance in level 3 = 80/1.55 = 52 ppm (mg/kg).
Toxic substance level 4 = 0.9(0.25*52) = 11.7 mg = 0.01X4 or X4 = 1,170 mg. The concentration of toxic substance in level 4 = 1,170/1.2 = 975 ppm (mg/kg).
The biomagnification series: 8.7 → 15 → 52 → 975 ppm (mg/kg).

2.5 Cycling in Ecosystems and in the Ecosphere

The renewable resources are used by man at a rate that in most cases is higher than the rate at which the resources are regenerated. It implies that the renewable resources are decreasing. The nonrenewable resources are used by man at a rate that is higher than the rate at which alternatives to the nonrenewable resources are found. The decreasing renewable and nonrenewable resources demonstrate that the earth is not in a sustainable development. Sustainable development is used in the same sense as in the Brundtland Report: a sustainable development means that we will hand over the earth with the same possibilities for the next generation to plan and live their life as the previous generations have given us when the earth was handed over to our generation.

Ecosystems are not using nonrenewable resources, but recycle the elements that are applied in the ecosystems. This does not imply that an ecosystem always has resources that are needed for growth, as has been discussed in Section 2.3. However, the ecosystems will in these cases adjust the rate of consumption to what is possible on a long-term basis, while the recycling will continue.

Some nonrenewable resources are recycled by man, for instance, iron, but a 100% recycling is not possible and therefore the nonrenewable resources are declining, although at a lower rate corresponding to the recycling.

The above 20–25 elements used by the organisms of ecosystems are recycled in the ecosystems, and the 6 elements that are considered absolutely necessary for life on the earth—C, H, O, N, P, and S—are of course particularly important for nature to recycle. Figure 2.8 gives two examples: phosphorus and nitrogen cycling in a lake. The two examples include the inorganic form of either inorganic reactive phosphorus or ammonium and nitrate, phosphorus, and nitrogen in phytoplankton, zooplankton, fish, sediment, and detritus. The phosphorus cycle includes the phosphorus in the pore water of the sediment. The detritus is mineralized to close the cycle, but the sediment can also release inorganic forms of phosphorus as phosphate and of nitrogen as ammonium. The cycles can include more or fewer details. In the case shown in Figure 2.8 the food chain is represented by nutrients, phytoplankton, zooplankton, and fish only. Phytoplankton could be represented by different groups of phytoplankton, as, for instance, nitrogen-fixing species, diatoms, and so on. The possible nitrogen fixing is indicated on the diagram, and so are the possibilities for denitrifications. The two diagrams for the phosphorus and nitrogen cycle are almost parallel. The process numbers on the phosphorus refer to the processes, which are the same for the two cycles. The following processes are included: (1) uptake of nutrient, (2) photosynthesis (solar radiation converted to the organic matter in phytoplankton), (3) grazing, (4) loss of feces to detritus by the grazing process, (5) predation of zooplankton by fish, (6) loss of feces to detritus by the predation, (7) settling of feces, (8) mineralization of detritus, (9) settling of phytoplankton and detritus, (10) fishery, (11) mineralization taking place in the sediment, (12) diffusion, (13–15 exchange) of reactive phosphorus, phytoplankton, and detritus with the environment (inflows and outflows), and (16–18) mortality of phytoplankton, zooplankton, and fish. Notice that the nitrogen cycle includes nitrification, which is the oxidation of ammonium to nitrate.

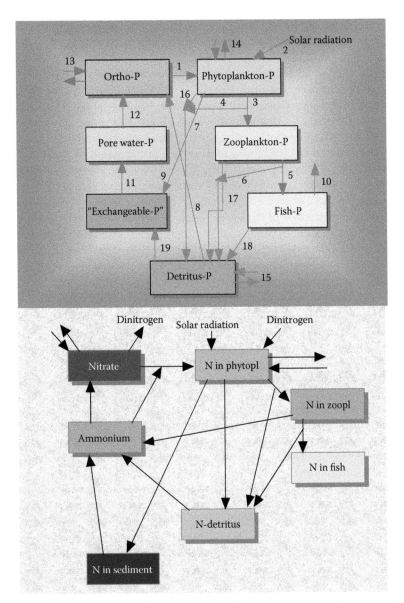

FIGURE 2.8   (See color insert.) The phosphorus and nitrogen cycle in a lake, using a simple food chain (nutrient-phytoplankton-zooplankton-fish) to describe the processes of the cycle. The cycles are closed by mineralization of detritus and by release of nutrient by the sediment. (From Jørgensen, S.E., and Bendoricchio, G., *Fundamentals of Ecological Modelling*, 3rd ed., Elsevier, Amsterdam, 2001.)

The elements in the entire ecosphere are cycling. It is important for a sustainable development of the earth that there are no crucial changes of the amounts of elements in the compartments making up the cycles. The global carbon and nitrogen cycles can be shown as in Figure 2.7 and Figure 2.8. The imbalance in the global carbon cycle is the accumulation of carbon dioxide in the atmosphere, because we are exhausting the storage of fossil fuels much faster than new fossil fuels can be formed. This entails what is known as the greenhouse effect, which can change the climate considerably, probably on the order of a 2 to 5°C increase of the temperature, dependent on the energy policy, toward the end of this century. The global nitrogen cycle is imbalanced by a huge production mainly from atmospheric nitrogen of nitrogen fertilizers, which are used in the lithosphere but by drainage water transported to the hydrosphere, where it causes increased eutrophication (too high primary production causing several water quality problems).

The imbalance of the global carbon cycle can only be solved effectively by cutting the consumption of fossil fuel by the development of alternative renewable energy sources, such as wind and solar energy. The imbalance of the nitrogen cycle can be solved either by a decreased production of fertilizers, which is hardly possible in a world with an increasing population, or by preventing the nitrogen from reaching the hydrosphere by treatment of wastewater and drainage water. The use of constructed or natural wetland ecosystems offers a good cost-moderate solution, particularly to treat drainage water. The nitrate and ammonium are by these treatment methods transformed into dinitrogen, which is transported to the atmosphere. A pattern of different types of wetlands in the landscape can reduce considerably the nitrogen loss to the hydrosphere.

All imbalances of global cycles or ecosystem cycles require a solution that respects the conservation principles. Matter or energy cannot be destroyed (or created) but only transformed, which implies that the energy sources used should be replaced by other energy forms that do not create imbalances of the spheres. This means that fossil fuel that causes accumulation of carbon dioxide in the atmosphere should be replaced by other energy forms, and nitrogen in wastewater and drainage water should be transformed to dinitrogen, which would be harmless in the atmosphere, as it contains 98% dinitrogen.

The appendix gives the composition of the spheres, which should be maintained at approximately the same level in the future to avoid imbalances of the global cycles.

2.6 Energy Flows in Ecosystems

Ecological models focus on the energy or mass flows in the ecosystem, because these flows determine the further development of the system and characterize the present conditions of the ecosystem. The pathway of energy flow may be compared with the usual electrical current in a wire with the driving force, X, balanced by a frictional force that develops almost in proportion to the rate of flow, J, so that there is a balance of forces:

$$X = R^*J \qquad (2.35)$$

where R is the resistance. Equation (2.35) is, as seen, a parallel to Ohm's law. $L = 1/R$ may be denoted conductivity, and Equation (2.35) may be reformulated to Onsager (1931), who stated that $L_{ij} = L_{ji}$ means that the conductivity from j to i is the same as the conductivity from i to j:

$$J = L^*X \qquad (2.36)$$

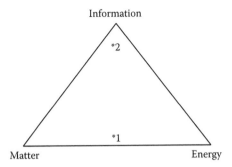

FIGURE 2.9 The conceptual triangle of matter, energy, and information. Point 1 corresponds to input of matter and energy, e.g., organic matter in the form of detritus. Point 2 corresponds to immigration of new species carrying genes and thereby information. It is accompanied by an input of matter and energy, but as shown in the diagram, it is relatively minor.

Ecological processes may be described in the same manner, for instance, the metabolism J (a flow rate) of a population N (a force driving the rate of metabolism):

$$J = L^*N \qquad (2.37)$$

The ecosystem also has a flow of mass under the driving influence of a thermodynamic force. The flux is the flow of food through a food chain, expressed in units such as carbon per square meter of ecosystem area per unit of time. The force is some function of the concentration gradient of organic matter and biomass.

The developments and reactions of ecosystems in general are, however, not only a question of the energy flow, which is in focus many times in this volume. Matter and information also play a major role. A conceptual triangle of matter, energy, and information is shown in Figure 2.9. No transfer of energy is possible without matter and information, and no matter can be transferred without energy and information. This implies that not only matter recycles but also energy and information recycle through the ecological network. The relationships between energy and information will be discussed in more detail later in this chapter. The higher the levels of information, the higher the utilization of matter and energy for further development of ecosystems away from the thermodynamic equilibrium; see Chapters 6 and 7.

E.P. Odum has described the development of ecosystems from the initial stage to the mature stage as a result of continuous use of the self-design ability (E.P. Odum, 1969, 1971). See the significant differences between the two types of systems listed in Table 2.5, and notice that the major differences are on the level of information. The content of information increases in the course of an ecological development, because an ecosystem encompasses an integration of all the modifications that are imposed on the environment. Thus, it is on the background of genetic information that systems develop that allow interaction of information with the environment. Herein lies the importance of the feedback from the organism to the environment; that means that an organism can only evolve in an evolving environment. The differences between the

TABLE 2.5   Differences between Initial Stage and Mature Stage Are Indicated (Odum, 1969)

Properties	Early Stages	Late or Mature Stage
A. Energetic		
P/R	>>1 <<1	Close to 1
P/B	High	Low
Yield	High	Low
Specific entropy	High	Low
Entropy production per unit of time	Low	High
Exergy	Low	High
Information	Low	High
B. Structure		
Total biomass	Small	Large
Inorganic nutrients	Extrabiotic	Intrabiotic
Diversity, ecological	Low	High
Diversity, biological	Low	High
Patterns	Poorly organized	Well organized
Niche specialization	Broad	Narrow
Size of organisms	Small	Large
Life cycles	Simple	Complex
Mineral cycles	Open	Closed
Nutrient exchange rate	Rapid	Slow
Life span	Short	Long
C. Selection and Homeostasis		
Internal symbiosis	Undeveloped	Developed
Stability (resistance to external perturbations)	Poor	Good
Ecological buffer capacity	Low	High
Feedback control	Poor	Good
Growth form	Rapid growth	Feedback controlled growth
Types	r-strategists	K-strategists

two stages include entropy and exergy. This latter concept will be discussed in the next chapter, and entropy has previously been defined as the extensive variable of the energy form heat.

Figure 2.10 illustrates the difference in energy utilization between the early stage and the mature stage. The biomass is smaller in the early stage, which implies that it can capture less solar radiation, while on the other hand it also requires less energy for maintenance (respiration). The mature stage captures, in contrast, more solar radiation but requires more energy for maintenance. In both cases a part of the solar radiation will be reflected.

The conservation laws of energy and matter set limits to the further development of "pure" energy and matter, while information may be amplified (almost) without

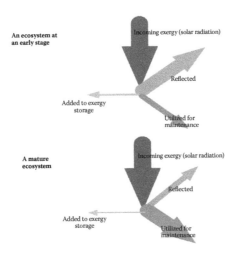

FIGURE 2.10 (See color insert.) The utilization of energy is compared for the early and mature stages. The width of the arrows gives the approximate relative amount of energy. The reflected sunlight is less for the mature stage, but also less of the incoming solar radiation is utilized to build new biomass, which is indicated as "added to exergy storage" (exergy and eco-exergy will be explained in Chapter 3, but consider them measures of the amount of solar energy that is used for growth). The maintenance requires, of course, more energy for the mature stage, because it contains more biomass, which implies more energy is used in respiration.

limits. These limitations lead to the concepts of limiting factors, which are playing a significant role in ecology, including systems ecology (see Section 2.3). Patten et al. (1997) have speculated what a world without the conservation principles would look like. Things would behave erratically. Something could arise from nothing. Mathematical counting would be meaningless. They conclude that if there is a scientific law more fundamental than the rest, it is probably the conservation principles of matter and energy.

All the attributes in Table 2.5 can be classified under three labels:

1. Growth of biomass
2. Growth of the ecological network
3. Growth of information

We will later return to this classification of the attributes, because it is important to be able to describe the growth and development possibilities by use of proper thermodynamic functions to capture the essence of the ecosystem growth and development, and thereby the reactions to perturbations of ecosystems, which is the focus of system ecology.

Example 2.4

Indicate for each of the three growth forms at least three of Odum's attributes that are covered by the labels. Indicate also three attributes that are covered by two or three growth forms simultaneously.

Solution

1. Growth of biomass includes P/R, P/B, yield, total biomass, inorganic nutrients, and entropy production per unit of time.
2. Growth of the ecological network includes patterns, niche specialization, life cycles, mineral cycles, and feedback control.
3. Growth of information includes information, ecological and biological diversity, and life span.

Exergy, internal symbiosis, type, buffer capacities, stability, and size of organisms cover two or three growth forms.

A major design principle observed in natural systems is the feedback of energy from storages to stimulate the inflow pathways as a reward from receiver storage to the inflow source (H.T. Odum, 1971). By this feature the flow values developed reinforce the processes that are doing useful work. Feedback allows the circuit to learn. A wider use of the self-organization ability of ecosystems in environmental or rather ecological management has been proposed by Odum (1988).

2.7 Summary of Important Points in Chapter 2

1. The first law of thermodynamics, which we will use to cover both energy and mass, because we do not consider radioactive processes in ecology, is used repeatedly in ecology to follow the transfer of mass and energy in ecosystems and the ecological processes and reactions. Mass and energy are conserved.
2. The conservation principle explains why an element or eventually another factor or important compound, for instance, water, may be limiting for growth. Living organisms need a combination of various elements and compounds to grow, and the element or compound that is used up first will inevitably stop the growth.
3. Ecosystems conserve matter and energy, which means that it is possible by bookkeeping of the exchange processes between the system and the environment to calculate the system's gain or loss of energy and matter. The bookkeeping can furthermore be used for each element, as we do not consider in ecosystems transformations of elements, like in nuclear physics.
4. Most chemical processes in the laboratory, ecosystems, or organisms are carried out at constant temperature and pressure. Gibbs free energy, G, covers the work at constant temperature and pressure. This implies that $dG = -dW$. At thermodynamic equilibrium, all gradients are eliminated according to the definition. This means that no work can be performed.

5. Standard free energies of formation of chemical compounds, $\Delta G°$, are very important for calculation of chemical equilibrium. The standard free energy of formation of a compound is the free energy of reaction by which it is formed from its elements, when all the reactants and products are in the standard state. The standard state can be used at 20°C and 1 atm pressure, conditions we often find in ecosystems. Free energy equations can be added and subtracted just as thermochemical equations. This implies that the free energy of any reaction can be calculated from the sum of the free energies of the products minus the sum of the free energies of the reactants:

$$\Delta G° = \Sigma\, G° \text{ (products)} - \Sigma\, G° \text{ (reactants)}$$

6. Free energy describes the chemical affinity under conditions of constant temperature and pressure: $\Delta G = G(\text{products}) - G(\text{reactants})$. When the free energy is 0 there is no net work obtainable by any change or reaction at constant temperature and pressure. The system is in a state of equilibrium. When the chemical energy change is positive for a proposed process, net work energy (free energy) must be put into the system to effect the reaction; otherwise it cannot take place. When the free energy change is negative, the reaction can proceed spontaneously by providing useful net work, which is used elsewhere in the system and its environment.

7. It can be shown for ideal gas reactions that the equilibrium constant K_p for the considered process is related to $-\Delta G°$ by the following equation: $-\Delta G° = RT \ln K_p$.

8. The relationship between growth and concentration for the needed elements and other factors and compounds can be expressed by the Michaelis-Menten equation, for instance:

$$\text{Growth} = \mu\text{max (PS/(PS + kp))}$$

or by two limiting factors:

$$\text{Growth} = \mu\text{max*min (NS/(kn + NS), (PS/(PS + kp))}$$

9. Different forms of energy and their intensive and extensive variables:

Energy Form	Extensive Variable	Intensive Variable
Heat	Entropy (J/K)	Temperature (K)
Expansion	Volume (m³)	Pressure (Pa = kg/s²m)
Chemical	Moles (M)	Chemical potential (J/mole)
Electrical	Charge (amp s)	Voltage (V)
Potential	Mass (kg)	(Gravity) (Height) (m²/s²)
Kinetic	Mass (kg)	0.5 (Velocity)² (m²/s²)

Note: Potential and kinetic energy is denoted mechanical energy.

10. The assimilation of toxic substances may be very different from the assimilation of food in general, which, as indicated, is about 67%. Only about 5–10% of heavy metals are assimilated, while the assimilation of an organic toxic substance with a high K_{ow} (the ratio of the solubility in octanol to the solubility in water for the considered organic substance), such as, for instance, DDT and PCB, is 90% or more. This can explain the bioaccumulation and the biomagnification. They can both be calculated by use of the conservation principle.

11. Ecosystems are not using nonrenewable resources, but recycle the elements that are applied in the ecosystems. Ecosystems recycle matter, energy, and information.

Exercises/Problems

1. A lake receives 300,000 m³ agricultural drainage water with 11 mg/L nitrogen and 0.2 mg/L phosphorus. The lake also receives 150,000 m³ wastewater with 30 mg/L nitrogen and 10 mg/L phosphorus. The lake is eutrophic and the phytoplankton has seven times as much nitrogen concentration as phosphorus concentration. How effective does the removal of phosphorus from the wastewater have to be to make phosphorus the limiting element?

2. What are the extensive variables for the following energy forms: electrical energy, chemical energy, and heat energy? What are the intensive variables for the following energy forms: pressure energy, heat energy, and kinetic energy?

3. A food chain consists of four levels. For each of the four levels the fraction of food (dry matter basis) assimilated can be considered to be two-thirds of the food intake. The amounts of food intake for the four levels per unit of time are, respectively, 20, 3, 1, and 0.3, expressed in kg/24 h. The respiration is 20% of the biomass per 24 h. The food for the first trophic level contains 3 ppm of a toxic substance that is assimilated 85% by all four trophic levels. The toxic substance is excreted by a rate of 0.005 per unit of time for all four trophic levels. Calculate the concentrations of the toxic substance in the four levels at steady state and indicate the series of biomagnification.

4. Explain how ecosystems are able to recycle trace metals.

5. 1.372 g of urea is burned in a bomb calorimeter with a net heat capacity of 5.00 kJ/K. A temperature rise of 2.95 K is observed. What volume change will occur if the combustion of 1 mole of urea was carried out at constant pressure of 1 atm. And what is the value of the enthalpy change in kJ/mole.

6. Blue-green algae are able to fix nitrogen. The process is a reaction between nitrogen and water, whereby ammonia and oxygen are formed. Calculate the standard free energy and the equilibirum constant for this process, when it can be found that the standard free energy of formation of 1 mole of water (gas) is 228.6 kJ, and of 1 mole of ammonia (gas), 16.6 kJ. The standard free energy of element is, according to definition, 0.

7. The energy of protein catabolism in mammals is lower than the energy of protein combustion in a calorimeter. Why?

8. What is the free energy of formation of carbon dioxide in the atmosphere at its normal sea level partial pressure of 0.00039 atm and a temperature of 298 K? The standard free energy of formation is 394.4 kJ.

3

Ecosystems: Growth and Development

Disorder is the information that you don't have.
Order is the information you have.

Ecosystems grow and develop, but under the constraints of the thermodynamic laws and the biochemistry of organisms. Ecosystems have three growth forms: growth of biomass, growth of the network, and growth of information, which cover Odum's attributes for the development of ecosystems. An understanding of how ecosystems are able to grow and develop in spite of the constraints requires an introduction to new thermodynamic concepts: maximum power, emergy, and exergy (both technological and ecological exergy, denoted eco-exergy).

3.1 The Maximum Power Principle

Lotka (1956) formulated the maximum power principle.

He suggested that systems prevail that develop designs that maximize the flow of *useful* (for maintenance and growth) energy, and H.T. Odum uses this principle to explain much about the structure and processes of ecosystems (Odum and Pinkerton, 1955). Similarly, Schrödinger (1944) pointed out that organization is maintained by extracting order from the environment. Boltzmann (1905) said that the struggle for existence is a struggle for free energy available for work, which is a definition very close to the maximum exergy principle, which will be introduced later in this chapter. Boltzmann's principle may be interpreted as the systems that are able to gain most free energy under the given conditions, i.e., to move most away from the thermodynamic equilibrium, will prevail. Such systems will gain the most biogeochemical energy available for doing work, and therefore have the most work capacity (storage of energy that can do work) to be able to struggle for their existence. This formulation is close to the tentative ecological law of thermodynamics (ELT) (see Chapters 6 and 7).

Power for electrical current is the product of voltage and current. Similarly, the product of J and X, see Equations (2.35) and (2.36), is power (Odum, 1983).

The organic matter accumulated in the biomass of an ecosystem may be defined as the ecopotential, E, equal to the free energy difference, that can be "released" by the process, ΔG, per unit of carbon, C. Thus, the ecopotential is a function of the concentration of biomass and organic matter.

The product of ecopotential and ecoflux, dC/dt, has the dimensions of power:

$$\text{Power} = E^*J = \Sigma\ \Delta G^* dC/(C^* dt) \tag{3.1}$$

where C is the concentration of biomass, measured as carbon. Power, as seen, is the increase in biomass concentration per unit of time converted to free energy.

Notice that the maximum power principle focuses on a rate, in Equation (3.1) indicated as dC/dt, the ecoflow, multiplied by the fraction that is able to do useful work, i.e., ΔG/C. Maximum power of an ecosystem becomes thereby equal to the sum of flow rates of useful energy in the system.

Odum (1983) defines the maximum power principle as a maximization of *useful* power. This implies that Equation (3.1) is applied to the ecosystem level by summing up all the contributions to the *total* power that are useful. This means that nonuseful power is not included in the summation. The difference between useful and nonuseful power will be further discussed below and in Chapters 6 and 7, because the emphasis on useful power is perhaps the key to understanding Odum's principle and utilizing it to interpret ecosystem properties.

Brown et al. (1993) and Brown (1995) have restated the maximum power principle in more biological terms. According to the restatement it is the transformation of energy into work (consistent with the term *useful power*) that determines success and fitness. Many ecologists have incorrectly assumed that natural selection tends to increase efficiency. If this were true, endothermic processes could never have evolved. Endothermic birds and mammals are extremely inefficient compared with reptiles and amphibians. They expend energy at high rates in order to maintain a high, constant body temperature, which gives high levels of activities independent of environmental temperature (Turner, 1970), which has been a clear advantage in the struggle for survival.

Brown (1995) defines fitness as reproductive power, dW/dt, the rate at which energy can be transformed into work to produce offspring. This interpretation of the maximum power principle is more consistent with the maximum exergy principle—the above-mentioned ELT, which will be presented in detail in Chapters 6 and 7—than with Lotka's and Odum's original idea.

In the book *Maximum Power—The Ideas and Applications of H.T. Odum*, Hall (1995) presented a clear interpretation of the maximum power principle, as it has been applied in ecology by H.T. Odum. The principle claims that power or output of useful work is maximized—not the efficiency and not the rate, but the trade-off between a high rate and high efficiency, yielding most useful energy = useful work. It is illustrated in Figure 3.1.

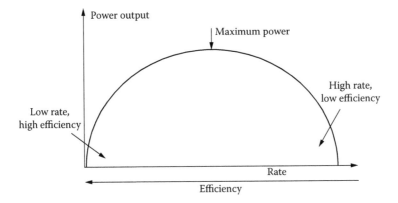

FIGURE 3.1 The maximum power principle claims that the development of an ecosystem is a trade-off (a compromise) between the rate and the efficiency, i.e., the maximum power output per unit of time.

Hall uses an interesting seminatural experiment by Warren (1970) to illustrate the application of the principle in ecology. Streams were stocked with different levels of predatory cutthroat trout. When predator density was low, there was considerable invertebrate food per predator, and the fish used relatively little energy in searching for food per unit of food obtained. With a higher fish-stocking rate, food became less available per fish, and each fish had to use more energy searching for it. Maximum production occurred at intermediate fish-stocking rates, which means intermediate rates at which the fish were able to utilize their food.

Hall (1995) mentions another example. Deciduous forests in moist and wet climates tend to have a leaf area index (LAI = leaf area per unit of surface area) of about 6. Such an index is predicted from the maximum power hypothesis applied to the net energy derived from photosynthesis. Higher leaf area index values produce more biomass by photosynthesis, but do so less efficiently because of the respirational demand of the additional leaf area. The respiration is proportional to the volume, which is approximately proportional to the leaf area in the exponent 1.5. A lower leaf area draws less power than the observed intermediate values of roughly 6.

Example 3.1

The photosynthetic production of biomass for a plant (photo) is with good approximation proportional (the coefficient is indicated as p = 0.7, unit 1/24 h) to the leaf area, A, as it determines the amount of solar radiation captured by the plant. This means photo = pA. Notice A, not LAI, is used. The respiration (coefficient r = 0.2, unit 1/(24 h m))—the energy needed for the maintenance of the leaf structure (which is far from thermodynamic equilibrium) (resp)—is proportional to the volume, which is A^1.5. Therefore, resp 0.2A^1.5. Set up a differential equation to find the optimum leaf area, AO. Develop also a STELLA model to follow the growth of A.

Solution

$$dA/dt = pA - rA^{1.5}$$

AO corresponds to $dA/dt = 0$ or $pA = rA^{1.5}$. This means that $AO = (p/r)^2$.

$$AO = 12.25. \ m^2$$

The STELLA diagram:

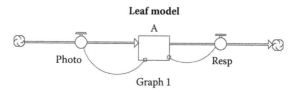

Leaf model

A

Photo   Resp

Graph 1

The numerical solution, $AO = 12.25 \ m^2$, corresponds to the steady-state value for A on the graph. LAI = 6 (LAI takes the shading effect into account) corresponds to one plant with a leaf area of $12.25 \ m^2$ per $2.04 \ m^2$ surface area, which seems reasonable.

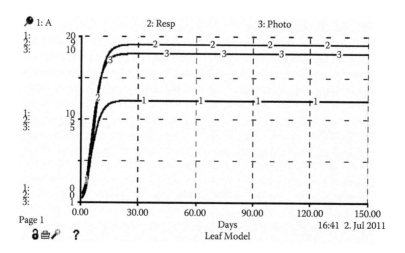

A larger leaf area would capture more sunlight, but the maintenance of the structure, the leaves, would require more energy than captured. On the other side, a smaller leaf area would cost less in maintenance but could capture less solar radiation. AO = AO = $12.25 \ m^2$ is the compromise between the two possibilities.

According to Gilliland (1982) and Andresen (1983), the same concept applies for regular fossil fuel power generation. The upper limit of efficiency for any thermal machine such as a turbine is determined by the Carnot efficiency. A steam turbine could run at 80% efficiency, but it would need to operate at a nearly infinitely slow rate. Obviously, we are not interested in a machine that generates revenues infinitely slowly, no matter how efficiently. Actual operating efficiencies for modern steam-powered generators are therefore closer to 40%, roughly half the Carnot efficiency.

The examples show that the maximum power principle is embedded in the irreversibility of the world, which is the core topic of the next chapter. The highest process efficiency can be obtained by endo-reversible conditions, meaning that all irreversibilities are located in the coupling of the system to its surroundings; there are no internal irreversibilities. Such systems will, however, operate too slowly. Power is zero for any endo-reversible system. If we want to increase the process rate, this implies that we also increase the irreversibility, and thereby decrease the efficiency. The maximum power is the compromise between endo-reversible processes and very fast, completely irreversible processes.

The power for ecosystems can be found by adding the flows of useful energy, which can be obtained from an ecological model describing the energy flows (or matter flows) in the ecosystem's ecological network. Usually, we have in our ecological models the flow rates, although they are generally indicated with less certainty than the state variables.

Example 3.2

Calculate the power corresponding to the P-cycling in Figure 2.8. P is considered to be 1% of the organic matter, and the following values are valid for the P-transfer processes in the lake: (unit mg/L 24 h)

1: 1.2; 3 + 4: 0.8; 5 + 6: 0.05; 7:2; 8: 0.5; 9: 2; 11: 1.1; 12: 0.2; 16: 2.4; 17: 0.2; 18: 0.02; 19: 1.25

The free energy of organic matter 18.7 kJ/g.

Solution

The power per liter of water is therefore

$(100 \cdot 1.2 + 2(0.8 + 0.05) + 2.8 + 0.5 + 2.0 + 1.1 + 0.2 + 2.4 + 0.2 + 0.02 + 1.25)18.7$
J/L 24h = 25.0 kJ/L 24h

3.2 Embodied Energy/Emergy

This concept was introduced by Odum (1983) and attempts to account for the energy required in formation of organisms in different trophic levels. The idea is to correct energy flows for their quality. Energies of different types are converted into equivalents

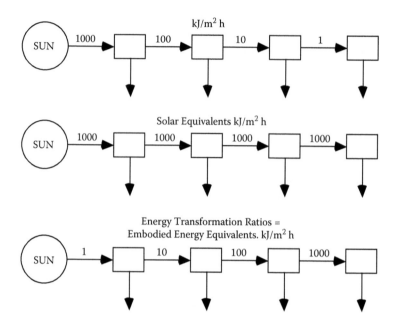

FIGURE 3.2 Energy flow, solar equivalents, and energy transformation ratios = embodied energy equivalents in a food chain. In the figure is presumed a linear food chain, while in reality it is a food web, which is considered in Table 3.1. (From Jørgensen, S.E., *Integration of Ecosystem Theories: A Pattern*, Kluwer, Dordrecht, 2002.)

of the same type by multiplying by the energy transformation ratio. For example, fish, zooplankton, and phytoplankton can be compared by multiplying their actual energy content by their solar energy transformation ratios. The more transformation steps there are between two kinds of energy, the greater the quality and the greater the solar energy required to produce a unit of energy (J) of that type. When one calculates the energy of one type, that generates a flow of another; this is sometimes referred to as the embodied energy of that type.

Figure 3.2 presents the concept of embodied energy in a hierarchical chain of energy transformation, and Table 3.1 gives embodied energy equivalents for various types of energy. Odum (1983) reasoned that surviving systems develop designs that receive as much energy amplifier action as possible. The energy amplifier ratio is defined in Figure 3.3 as the ratio of output B to the control flow C. Odum (1983) suggested that in surviving systems the amplifier effects are proportional to embodied energy, but full empirical testing of this theory still needs to be carried out in the future.

One of the properties of high-quality energies is their flexibility. Whereas low-quality products tend to be special, requiring special uses, the higher-quality part of a web is of a form that can be fed back as an amplifier to many different units throughout the web. For example, the biochemistry at the bottom of the food chain in algae and microbes is diverse and specialized, whereas the biochemistry of top animal consumer units tends

TABLE 3.1   Embodied Energy Equivalents for Various Types of Energy

Type of Energy/Item or Transformity (seJ/J)	Embodied Energy Equivalents
Solar energy	1.0
Winds	315
Gross photosynthesis	920
Coal	6,800
Tide	11,560
Electricity	27,200
Kinetic energy of spring flow	7,170
Detritus	6,600
Gross plant production	1,620
Net plant production	4,660
Herbivores	127,000
Carnivores	4,090,000
Top carnivores	40,600,000

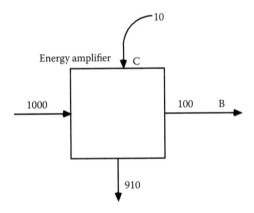

FIGURE 3.3   The energy amplifier ratio, R, is defined as the ratio of output B to control flow C. This means that R = 10 in this case. (From Jørgensen, S.E., *Integration of Ecosystem Theories: A Pattern*, Kluwer, Dordrecht, 2002.)

to be similar and general, with services, recycles, and chemical compositions usable throughout.

Hannon (1973, 1979, 1982) and Herendeen (1981) applied energy intensity coefficients as the ratios of assigned embodied energy to actual energy to compare systems with different efficiencies.

The difference between embodied energy flows and power simply seems to be a conversion to solar energy equivalents of the free energy. The increase in biomass, in Equation (3.1), is a conversion to the free energy flow, and the definition of embodied energy is a further conversion to solar energy equivalents.

Embodied energy is, as seen from these definitions, determined by the biogeochemical energy *flow* into an ecosystem component, measured in solar energy equivalents. The stored emergy, Em, per unit of area or volume to be distinguished from the emergy flows can be found from

$$Em = \sum_{i=1}^{i=n} Q_i * c_i \qquad (3.2)$$

where Q_i is the quality factor = transformity (seJ/J), which is the conversion to solar equivalents, as illustrated in Table 3.1 and Figure 3.2, and c_i is the concentration expressed per unit of area or volume.

The calculations by Equation (3.2) reduce the difference between stored emergy (= embodied energy) and stored exergy, another thermodynamic concept, which as will be shown later in the chapter, also can be found with good approximations as the sum of concentrations times a quality factor, to a difference in the definition of the quality factor. The quality factor for eco-exergy (detailed definition of exergy and eco-exergy; see below) accounts for the information embodied in the various components in the system (detailed information is given below and in Chapter 6), while the quality factor for emergy accounts for how much solar energy it costs to form the various components. Emergy thereby calculates how much solar energy (which is our ultimate energy resource) it costs to obtain one unit of biomass of various organisms, while eco-exergy accounts for how much "first class" energy (defined as energy that can do work, as will be presented in more detail below and in Chapter 6) the organisms have, as a result of the complex interactions in an ecosystem. Both concepts attempt to account for the quality of the energy: emergy by looking into the energy flows in the ecological network to express the energy costs in solar equivalents, and eco-exergy by considering the amount of information (which also contains first-class energy able to do work; see also Chapter 6) that the components have embodied. Both concepts can be used to understand ecosystems better and to reveal the energetic consequences of the ecological processes. Emergy calculations of products are also made and are used to assess how effectively various productions are able to utilize the ultimate energy source—sunlight. Comparisons of emergy balances for different countries and the emergy of their export and import can be used to find how much emergy—solar radiation—it costs to drive the different countries and how much they gain or lose in solar energy by the export and import. Generally, the industrialized countries are gaining by their exports/imports compared with the developing countries.

The differences between the two concepts, exergy and emergy, may be summarized as follows:

1. Emergy has no clear reference state, which is not needed, as it is a measure of energy flows, while exergy is defined relative to the environment (see also Section 4.8 and Chapter 5).
2. The quality factor of exergy is based on the content of information, while the quality factor for emergy is based on the cost in solar equivalents.
3. Exergy is anchored in thermodynamics (but not the classical thermodynamics) and has a wider theoretical basis.

4. The quality factor, Ω, may be different from ecosystem to ecosystem, and in principle, it is necessary to assess in each case the quality factor based on an energy flow analysis, which is sometimes cumbersome to make.

Nevertheless, emergy calculations are in many contexts better than exergy able to capture the ecological economy. In Chapter 15, it will be shown and discussed how the ratio of exergy to emergy actually is a very informative ecological indicator.

The quality factors listed in Table 3.1 or in Brown and Clanahan (1992) may be used generally as good approximations. The quality factors used for computation of exergy, ß (see later in this chapter and in Chapter 6), require a knowledge of the nonnonsense genes (information content) of various organisms, which sometimes is surprisingly difficult to assess. A number of eco-exergy quality factors have been found. They can, from a theoretical point of view, be used generally, but our knowledge about the information content of different organisms is unfortunately rather limited.

In his book *Environmental Accounting: Emergy and Environmental Decision Making*, Odum (1996) used calculations of emergy to estimate the sustainability of the economy of various countries. As emergy is based on the cost in solar equivalents, which is the only long-term available energy, it seems to be a sound first estimation of sustainability.

Example 3.3

The state variables of the P-cycle in Figure 2.8 have the following concentrations: Ortho P: 0.1 mg/L; Phyt-P: 0.1 mg/L; Zoopl-P: 0.02 mg/L; Fish-P: 0.0025 mg/L; Det-P: 0.025 mg/L; Exch. P: 0.022; and Por-P: 0.4 mg/L.

Calculate the emergy per liter of lake water when it is presumed that P for phytoplankton, zooplankton, detritus, fish, and exchangeable P is 1%.

The transformity for inorganic P is 0. It is presumed that organic matter has 18.7 kJ/g free energy.

Solution

Equation (3.2) and Table 3.1 can be used directly, as we know the transformities for the state variables.

(0.1*0 + 0.1*100*1,620 + 0.02*100*127,000 + 0.0025*100*4,090,000 + 0.025*100*6,600 + 0.022*100*6,600 + 0.4*0)*18.7 J/L = 24750 kJ/L

The high number is dominated by the high transformities of zooplankton and fish.

3.3 Ecosystem as a Biochemical Reactor

The fuel of ecosystems is organic matter, detritus. It is therefore relevant to calculate the free energy of dead organic matter. The chemical potential of dead organic matter, indexed $i = 1$, can be expressed from classical thermodynamics (e.g., Russel and Adebiyi, 1993) as

$$\mu_1 = \mu_1^o + RT \ln c_1/c_{1,o} \tag{3.3}$$

where μ_1 is the chemical potential. The difference $\mu_1 - \mu_1^o$ is known for detrital organic matter, which is a mixture of carbohydrates, fats, and proteins. Approximately 18.7 kJ/g may be applied for the free energy content of average detritus. Obviously, the value is higher for detritus originating from birds, as they, on average, contain more fat. Coal has a free energy content of about 30 kJ/g, and mineral oil 42 kJ/g. Both coal and mineral oil are concentrated forms of detritus from previous periods of the earth. c_1 is the concentration of the detritus in the considered ecosystem, and $c_{1,o}$ is the concentration of detritus in the same ecosystem but at thermodynamic equilibrium.

Generally, $c_{1,o}$ can be calculated from the definition of the probability, $P_{i,o}$, of finding component i at thermodynamic equilibrium, which is

$$P_{i,o} = c_{i,o} \Big/ \sum_{i=0}^{n} c_{i,o} \tag{3.4}$$

If this probability can be determined, then in effect the ratio of $c_{i,eq}$ to the total concentration is also determined. As the inorganic component, c_0, is very dominant at thermodynamic equilibrium, Equation (3.4) can be approximated as

$$P_{i,o} = c_{i,o}/c_{0,0} \tag{3.5}$$

By a combination of Equations (3.3) and (3.5), we get

$$P_{1,o} = [c_1/c_{0,o}] \exp [-(\mu 1 - \mu_1^o)/RT] \tag{3.6}$$

The equilibrium constant for the process describing the aerobic (presence of oxygen) decomposition of detritus at 300 K, can be found based upon the above-mentioned values. We could presume a molecular weight of about 100,000 (more accurately, 102,400; Morowitz, 1968) and with a typical composition of 3,500 carbons, 6,000 hydrogens, 300 oxygens, and 600 nitrogens:

$$C_{3,500}H_{6,000}O_{3,000}N_{600} + 4,350\ O_2 \rightarrow 3,500\ CO_2 + 2,700\ H_2O + 600\ NO_3^- + 600\ H+ \tag{3.7}$$

$$K = [CO_2]^{3,500}[NO_{3-}]^{600}[H+]^{600}/[C_{3,500}H_{6,000}O_{3,000}N_{600}]\ [O_2]^{4,350}$$

Since water is omitted from the expression of K. We have

$$-\Delta G = RT \ln K \tag{3.8}$$

$-\Delta G = -18.7$ kJ/g.104,400 g/mole = 1,952 MJ/mole = 8.2 J/mole.300 ln K

which implies that ln K = 793,496, or K is about $10^{344,998}$.

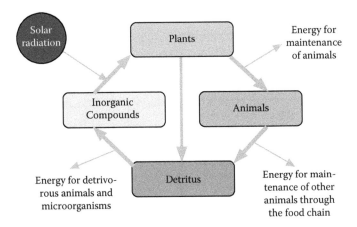

FIGURE 3.4 (See color insert.) An ecosystem is a biochemical reactor. The input of energy comes from the solar radiation. The biologically important elements cycle and carry the energy, which is utilized by heterotroph organisms to support the life processes.

The equilibrium constant is, in other words, enormous. The spontaneous formation of detritus in the form of a compound with the molecular weight of about 100,000 therefore has a very small probability.

Even if we consider detritus with a low molecular weight corresponding to detritus partially decomposed, the K-value is still very high. If we presume a 100 times smaller molecular weight, the exponent is 100 times smaller, or about 3,500—still a very high K-value. It is therefore understandable that detritus is decomposed spontaneously, and thereby yields energy to heterotroph organisms. The opposite process corresponds to what may be the result of photosynthesis, the conversion of solar radiation (energy) into chemical energy. This process is realized because the solar radiation delivers the needed energy and chlorophyll is an effective enzyme for the process.

Figure 3.4 shows the resulting biochemical reactions of an ecosystem, i.e., how the ecosystem works as a biochemical reactor. The biologically important elements are cycling and used again and again to build up biochemically important compounds as, for instance, proteins, lipids, and carbohydrates. These compounds are carrying the energy of the solar radiation and thereby supporting the maintenance of life and the cycling processes. The cycle may be compared with a Carnot cycle. The hot reservoir (the sun) delivers the energy, which is utilized to do work. The heat energy is delivered to the cold reservoir at the ambient temperature (the temperature of the earth). The work is after it has been used to transform to heat, which is also delivered to the environment. Figure 3.5 shows the cycling of matter in more detail.

Figure 2.1 illustrates the free energy (energy that can do work; see Section 2.2) of various ecologically important nitrogen compounds. Proteins with the highest free energy are the result of photosynthesis or of the biochemical synthesis in heterotroph organisms. They are important food items for heterotroph organisms as suppliers of the important building blocks, amino acids, and of energy.

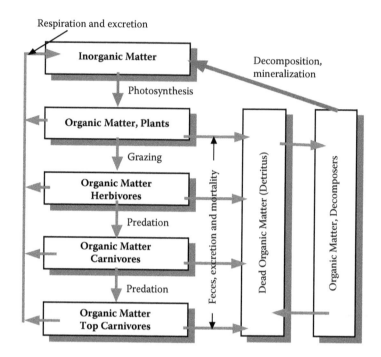

FIGURE 3.5 The biochemical cycling of matter in ecosystems. (From Jørgensen, S.E., *The Principles of Pollution Abatement*, Elsevier, Amsterdam, 2000.)

3.4 Technological and Ecological Interpretation of the Thermodynamic Concept Exergy

Previously in this chapter, we used the term *work capacity* to express the ability of a part of the total energy to perform work in contrast to heat energy at the temperature of the environment, that is, without work capacity. The classical thermodynamics uses the G-function (free energy) to cover the work capacity, but when we are dealing with distances very far from thermodynamic systems, we can no longer use state variables that are independent of the pathway. Furthermore, we need in different situations different reference states. Therefore, we have to define a work capacity that can also be used very far from thermodynamic equilibirum systems.

Exergy is defined as the amount of work (= entropy-free energy) a system can perform when it is brought into thermodynamic equilibrium with its environment (Jørgensen et al., 1999). Figure 3.6 illustrates the definition of exergy as it is used in technology, for instance, to find the work capacity of a power plant. The considered system is characterized by the extensive state variables S, U, V, N1, N2, N3, ..., where S is the entropy, U is the energy, V is the volume, and N1, N2, N3, ..., are the moles of various chemical compounds, and by the intensive state variables, T, p, μ_{c1}, μ_{c2}, μ_{c3}, ... The system is coupled to a reservoir, a reference state, by a shaft. The system and the reservoir form a closed system. The reservoir (the environment) is characterized by the intensive state variables

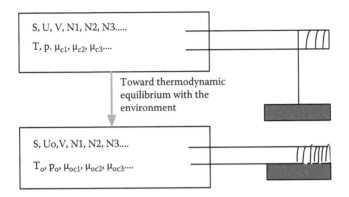

FIGURE 3.6 Definition of technological exergy is shown.

T_o, p_o, μ_{c1o}, μ_{c2o}, μ_{c3o}, ..., and as the system is small compared with the reservoir, the intensive state variables of the reservoir will not be changed by interactions between the system and the reservoir. The system develops toward thermodynamic equilibrium with the reservoir and is simultaneously able to release entropy-free energy to the reservoir. During this process the volume of the system is constant, as the entropy-free energy must be transferred through the shaft only.

The entropy is also constant as the process is an entropy-free energy transfer from the system to the reservoir, but the intensive state variables of the system become equal to the values for the reservoir. The total transfer of entropy-free energy in this case is the exergy of the system. It is seen from this definition that exergy is dependent on the state of the total system (= system + reservoir) and not dependent entirely on the state of the system. Exergy is therefore not a state variable. In accordance with the first law of thermodynamics, the increase of energy in the reservoir, ΔU, is $U - U_o$, where U_o is the energy content of the system after the transfer of work to the reservoir has taken place. According to the definition of exergy, Ex, we have:

$$Ex = \Delta U = U - U_o$$

$$as\ U = TS - pV + \sum_c \mu_c N_i \tag{3.9}$$

(when we only consider heat, spatial energy [displacement work]), and chemical energy, see any textbook in thermodynamics and Section 2.2), and correspondingly for U_o:

$$U_o = T_o S - p_o V + \sum_c \mu_{co} N_i \tag{3.10}$$

we get the following expression for exergy, excluding of course, in this case, kinetic energy, potential energy, electrical energy, radiation energy, and magnetic energy:

$$Ex = S\,(T - T_o) - V\,(p - p_o) + \sum_c (\mu_c - \mu_{co})N_i \tag{3.11}$$

These energy forms could, however, easily be included.

Notice that the above shown equation also emphasizes that exergy is dependent on the state of the environment (the reservoir = the reference state), as the exergy of the system is dependent on the intensive state variables of the reservoir. Notice furthermore that exergy is not conserved—only if entropy-free energy is transferred, which would imply that the transfer is reversible. All processes in reality are, however, irreversible, which means that exergy is lost (and entropy is produced). It will be mentioned in more detail in Chapter 4, where the second law of thermodynamics will be presented.

Energy is conserved by all processes according to the first law of thermodynamics. It is therefore wrong to discuss an energy efficiency of an energy transfer because it will always be 100%, while the exergy efficiency is of interest, because it will express the ratio of useful energy (work energy) to total energy, which is always less than 100% for real processes. All transfers of energy imply that exergy is lost because energy is transformed to heat at the temperature of the environment.

It is therefore of interest to set up for all environmental systems, in addition to an energy balance, an exergy balance. Our concern is loss of exergy, because that means loss of work capacity, or "first class energy," which can do work, is lost as "second class energy" (heat at the temperature of the environment), which cannot do work. So, the particular properties of heat, including that temperature is a measure of the movement of molecules, limit our possibilities to utilize this energy form to do work. The total amount of heat energy is the entropy times the absolute temperature, but the amount of heat energy that we can utilize to do work is only the high temperature that we may have provided in, for instance, steam generator minus the temperature of the environment times the available entropy = S ($T_g - T_{environment}$). Due to these limitations we have to distinguish between exergy, which can do work, and anergy, which cannot do work, and all real processes imply inevitably a loss of exergy as anergy (see also Chapter 4 for details about the second law of thermodynamics).

Exergy seems more useful to apply than entropy to describe the irreversibility of real processes, as it has the same unit as energy and is an energy form, while the definition of entropy is more difficult to relate to concepts associated with our usual description of the reality. In addition, entropy is not clearly defined for "far from thermodynamic equilibrium systems," particularly for living systems (see, for instance, Tiezzi, 2003). Moreover, it should be mentioned that the self-organizing abilities of systems are strongly dependent on the temperature, as discussed in Jørgensen et al. (1999). Exergy takes the temperature into consideration as the definition shows, while entropy doesn't. It implies that exergy at 0 K is 0 and at minimum. Negative entropy is sometimes used (see, for instance, Schrödinger, 1944), but it does not express the ability of the system to do work (or we may call it the "creativity" of the system, as creativity requires work). Exergy becomes a good measure of the creativity, which increases proportionally with the temperature. Notice also that in classical thermodynamics, entropy cannot be negative. Furthermore, exergy facilitates the understanding of the difference between low-entropy energy and high-entropy energy, as exergy is entropy-free energy.

Finally, notice that information contains exergy. Boltzmann (1905) showed that the free energy of the information that we actually possess (in contrast to the information we need to describe the system) is k*T*ln I, where I is the information we have about the state of the system, for instance, that the configuration is 1 out of W possible (i.e., W = I) and k is Boltzmann's constant = $1.3803*10^{-23}$ (J/molecules*deg).

This implies that one bit of information has the exergy equal to k T ln 2. Transformation of information from one system to another is often almost an entropy-free energy transfer. If the two systems have different temperature, the entropy lost by one system is not equal to the entropy gained by the other system, while the exergy lost by the first system is equal to the exergy transferred and equal to the exergy gained by the other system, provided that the transformation is not accompanied by any loss of exergy, which it always will be in reality. In this case, it is obviously more convenient to apply exergy than entropy.

Exergy of the system measures the contrast—it is the difference in free energy if there is no difference in pressure and temperature, as may be assumed for an ecosystem or an environmental system and its environment—against the surrounding environment. If the system is in equilibrium with the surrounding environment, the technological exergy is zero. The only way to move systems away from equilibrium is to perform work on them. Therefore, it is reasonable to use the available work, i.e., the exergy, as a measure of the distance from thermodynamic equilibrium.

Exergy for ecosystems presumes reference to the same system (ecosystem) but at thermodynamic equilibrium, which means that all the components are inorganic at the highest possible oxidation state and homogeneously distributed in the system (no gradients).

It is illustrated in Figure 3.7. As the chemical energy embodied in the organic components and the biological structure contribute far more to the exergy content of the system, there seems to be no reason to assume a (minor) temperature and pressure difference between the system and the reference environment. Under these circumstances we can calculate the exergy, which we will name eco-exergy to distinguish from the technological exergy defined above, as coming entirely from the chemical energy:

$$\sum_c (\mu_c - \mu_{co}) \, N_i \qquad (3.12)$$

This represents the nonflow chemical exergy. It is determined by the difference in chemical potential $(\mu_c - \mu_{co})$ between the ecosystem and the same system at thermodynamic equilibrium. This difference is determined by the concentrations of the considered components in the system and in the reference state (thermodynamic equilibrium), as is the case for all chemical processes. We can measure the concentrations in the ecosystem, but the concentrations in the reference state (thermodynamic equilibrium) can be based on the usual use of chemical equilibrium constants. If we have the process

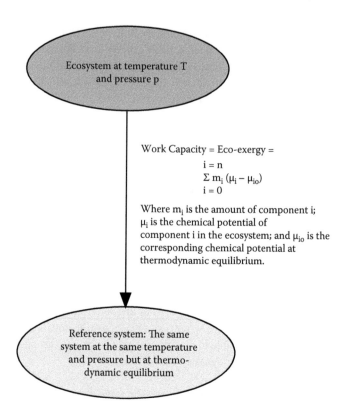

FIGURE 3.7 The exergy content of the system is calculated in the text for the system relative to a reference environment of the same system at the same temperature and pressure, but as an inorganic soup with no life, biological structure, information, or organic molecules.

$$\text{Component A} \leftrightarrow \text{inorganic decomposition products} \qquad (3.13)$$

it has a chemical equilibrium constant, K:

$$K = [\text{inorganic decomposition products}]/[\text{component A}] \qquad (3.14)$$

The concentration of component A at thermodynamic equilibrium is difficult to find, but we can find the concentration of component A at thermodynamic equilibrium from the probability of forming A from the inorganic components, as will be presented in the next section.

Eco-exergy is a concept close to Gibbs free energy, but opposite Gibbs free energy, eco-exergy has a different reference state from case to case (from ecosystem to ecosystem), and it can furthermore be used far from thermodynamic equilibrium, while Gibbs free energy in accordance to its exact thermodynamic definition is only a state function close to thermodynamic equilibrium.

We find by these calculations the eco-exergy of the system compared with the same system at the same temperature and pressure, but in form of an inorganic soup without any life, biological structure, information, or organic molecules. As $(\mu_c - \mu_{co})$ can be found from the definition of the chemical potential replacing activities by concentrations, we get the following expression for the eco-exergy (see Equation (2.21):

$$Ex = RT \sum_{i=0}^{i=n} C_i \ln C_i/C_{i,o} \qquad (3.15)$$

where R is the gas constant (8.317 J/K moles = 0.08207 L.atm/K moles), T is the temperature of the system, and C_i is the concentration of the i-th component expressed in a suitable unit; e.g., for phytoplankton in a lake C_i could be expressed as mg/L or as mg/L of a focal nutrient. $C_{i,o}$ is the concentration of the i-th component at thermodynamic equilibrium and n is the number of components. $C_{i,o}$ is, of course, a very small concentration (except for $i = 0$, which is considered to cover the inorganic compounds), corresponding to a very low probability of forming complex organic compounds spontaneously in an inorganic soup at thermodynamic equilibrium. $C_{i,o}$ is even lower for the various organisms, because the probability of forming the organisms is very low with their embodied information, which implies that the genetic code should be correct.

By using this particular exergy based on the same system at thermodynamic equilibrium as reference, the eco-exergy becomes dependent only on the chemical potential of the numerous biochemical components that are characteristic for life. This is consistent with Boltzmann's statement that life is a struggle for free energy.

The total eco-exergy of an ecosystem *cannot* be calculated exactly, as we cannot measure the concentrations of *all* the components or determine all possible contributions to the eco-exergy of an ecosystem. If we calculate the eco-exergy of a fox, for instance, the above shown calculations will only give the contributions coming from the biomass and the information embodied in the genes, but what is the contribution from the blood pressure, the sexual hormones, and so on? These properties are at least partially covered by the genes, but is that the entire story? We can calculate the contributions from the dominant components, for instance, by the use of a model or measurements that cover the most essential components for a focal problem. The *difference* in eco-exergy by *comparison* of two different possible structures (species composition) is here decisive. Moreover, eco-exergy computations always give only relative values, as the eco-exergy is calculated relative to the reference system.

Notice that the definition of eco-exergy is very close to free energy. Eco-exergy is, however, a *difference* in free energy between the system and the same system at thermodynamic equilibrium. The reference system used is different for every ecosystem according to the definition of eco-exergy. In addition, free energy is not a state function far from thermodynamic equilibrium. Consider, for instance, the immediate loss of free energy (or let us use the term eco-exergy, as already proposed, to make the use of the concepts more clear) when an organism dies. A microsecond before death the information can be used, and after death the information is worthless and should therefore not be included in the calculation of eco-exergy. Therefore, eco-exergy cannot be differentiated.

3.5 Eco-Exergy and Information

Biological structures maintain and reproduce themselves by transforming energy and information from one form to another. An important feature of life is that the information laid down in the genetic material is developed and transferred from one generation to the next. When biological materials are used to the benefit of mankind, it is in fact the organic structures and the information contained therein that are of advantage, for instance, when using wood.

In statistical mechanics, entropy is related to probability. A system can be characterized by averaging ensembles of microscopic states to yield the macrostate. If W is the number of microstates that will yield one particular macrostate, the probability P that this particular macrostate will occur as opposed to all other possible macrostates is proportional to W. It can further be shown that

$$S = k*\ln W \tag{3.16}$$

where k is Boltzmann's constant, $1.3803.10^{-23}$ J/(molecules.K). The entropy is a logarithmic function of W, and thus measures the total number of ways that a particular macrostate can be constituted microscopically. $kAv = R$, where Av is Avogadro's number. Notice that entropy measures the information we need.

S may be called thermodynamic information, meaning the amount of information *needed* to describe the system, which must *not* be interpreted as the information that we actually possess. The information that we have becomes equal to negentropy (–S), which from a strict thermodynamic point of view does not exist. According to the third law of thermodynamics, entropy cannot be negative, and it would be more correct to use the phrase the free energy (or exergy) of the information we have = kT ln W. Boltzmann also used the expression the free energy of information (Boltzmann, 1905). The more microstates there are and the more disordered they are, the more information is required and the more difficult it will be to give a complete description of the system—and the higher is the entropy. This means that entropy + information = constant. When we gain information, entropy decreases, and when we lose information, entropy increases. The constant expresses the total information needed to know all details of a considered system. When we have the full information about the considered system, information = constant and entropy = 0. When we have no knowledge about the system, the entropy = constant and information = 0.

If we consider that this book contains 10^6 signs selected among 40 possible signs, the entropy according to Equation (2.54) is $1.3803*10^{-23}.10^6$ ln $40 = 5.09.10^{-17}$ J/K. The entropy is the disorder resulting from 1,000,000 signs selected among 40 possible types. If, however, we understand the information embodied in the 1,000,000 signs, i.e., we can read English, the free energy (eco-exergy) of the information in this book at room temperature (about 300 K) becomes $1.527.10^{-14}$ J.

It is possible to distinguish between the contribution to the exergy of information and of biomass (Svirezhev, 1998). p_i defined as c_i/A, where

$$A = \sum_{i=1}^{n} c_i$$

is the total amount of matter in the system, is introduced as a new variable in Equation (3.15):

$$Ex = A\,RT \sum_{i=1}^{n} p_i \ln p_i/p_{io} + A \ln A/A_o \qquad (3.17)$$

As $A \approx A_o$, exergy becomes a product of the total biomass A (multiplied by RT) and Kullback measure:

$$K = \sum_{i=1}^{n} p_i \ln (p_i/p_{io}) \qquad (3.18)$$

where p_i and p_{io} are probability distributions, a posteriori and a priori to an observation of the molecular detail of the system. This means that K expresses the amount of information that is gained as a result of the observations. If we observe a system, which consists of two connected chambers, we expect the molecules to be equally distributed in the two chambers; i.e., $p_1 = p_2$ is equal to 1/2. If, on the other hand, we observe that all the molecules are in one chamber, we get $p_1 = 1$ and $p_2 = 0$.

Exergy density can therefore be found as biomass density times RTK.

Specific exergy is exergy relative to the biomass and for the i-th component: Sp. $ex._i$ = Ex_i/c_i. This implies that the total specific exergy per unit of area or per unit of volume of the ecosystem is equal to RTK.

Due to the incoming free energy of solar radiation, an ecosystem is able to move away from the thermodynamic equilibrium; i.e., the system evolves, obtains more information, and organization. The ecosystem must produce entropy for maintenance, but the low-entropy energy flowing through the system may be able to more than cover this production of disorder, resulting from maintenance of the order and structure. Chapter 6 will give more detailed information about the development of the ecosystem, if the free energy flow through the system is more than required for maintenance.

Boltzmann (1905) emphasized that competition in nature is a struggle for free energy and that information embodies free energy. By the use of chemical statistics Boltzmann expresses entropy as $S = R \ln W$, and free energy as $-RT \ln W$, where W is the number of microstates (that are presumed not known). If we presume that we know the present (a posteriori) probability, $p_i = 1.0$, while p_{io} (a priori) has a probability corresponding to W, it is possible to understand that eco-exergy = Biomass$\cdot$RTK is close to what Boltzmann

considered free energy. The difference is that free energy is not a state function for far—
or rather very far—from equilibrium systems as ecosystems, while we have accepted that
eco-exergy is not a state function, and that it is a relative measure of the work capacity
(relative to the ecosystem at thermodynamic equilibrium).

3.6 Summary of Important Points in Chapter 3

1. Growth and development of ecosystems have three forms, covering E.P.
   Odum's attributes:
   a. Growth of biomass
   b. Growth of the ecological network
   c. Growth of information
2. Power is maximized in ecosystems. It is equal to the sum of flow rates of useful
   energy in the system. It has the unit J/(L 24 h).
3. Emergy expresses the costs in solar energy and can be found as

$$\sum_{i=1}^{i=n} \Omega_i * c_i$$

   where Ω_i is the quality factor = transformity (seJ/J).
4. The equilibrium constant for high molecular organic matter is extremely big,
   which indicates that the decomposition has a very high affinity, while the opposite
   process (building high molecular organic matter) is very improbable.
5. Eco-exergy measures the distance from thermodynamic equilibrium, which
   equals the content of work capacity relative to the thermodynamic equilibrium.

   Exergy density can be found as biomass density times RTK.

6. Eco-exergy = –TS = kT ln W. Notice S measures the information we need to get,
   while Ex measures the work capacity of the information that we do have.

Exercises/Problems

1. Calculate the emergy per liter of an aquatic ecosystem, where phytoplankton has
   a concentration of 4 mg/L, zooplankton 1 mg/L, fish 0.2 mg/L, top carnivore fish
   0.05 mg/L, and detritus 5 mg/L.
2. Explain why formation of organic matter with a molecular weight of 10,000 can
   not be formed spontaneously? Calculate approximately which magnitude of concentration we should expect for organic matter of average composition in a system
   at thermodynamic equilibrium. We find high molecular organic matter in ecosystems. How is this possible?
3. Explain why eco-exergy measures the distance from thermodynamic equilibrium.
4. What is the difference between Gibbs free energy, technological exergy, and eco-
   exergy? Which of these three thermodyanmic variables are not state variables?
   Explain why.

5. Explain why S and eco-exergy have opposite signs. What is the difference between the information that we have and that which we need? Why is negative entropy not in accordance with classical thermodynamics?

6. Find the standard free energy change of ATP hydrolysis by using the following information:

Glucose + ATP = glucose-6-phosphate + ADP $DG_0 = -16.7$ kJ/mole

Glucose-6-phosphate = glucose + phosphate $DG_0 = -13.8$ kJ/mole

4

Irreversibility and Order: The Second and Third Laws of Thermodynamics

No use crying over spilt milk.
Only movies, realities are not reversible.

Ecosystems are open systems. They have to receive an inflow of free energy—energy that can do work—to be able to maintain their complex structure. In accordance with the second law of thermodynamics, all dynamic systems will inevitably produce entropy, lose order, and free energy. Ecosystems therefore need an inflow of energy that can do work to combat these consequences of the second law of thermodynamics. Ecosystems are, however, not only physical open but also ontic open. Due to the very high complexity of ecosystems, an uncertainty principle for ecological observations is valid.

The second and third laws of thermodynamics are introduced and their consequences for ecosystems presented. Ecosystem development and growth under the constraints of the thermodynamic laws can be well described by the use of eco-exergy. It is shown how it is possible to calculate eco-exergy for ecosystems.

4.1 Open Systems

Ecosystems are open systems in the sense that they are open for mass and energy transfers. Chapter 8 is devoted to the consequences and a quantification of this crucial property by ecosystems. This chapter will cover the concept in the light of the second law of thermodynamics, which is the core topic of this chapter.

Ecosystems receive energy from solar radiation and receive water from precipitation, dry deposition from the atmosphere, inputs by wind, and inflows and outflows of various types, plus emigration or immigration of species. A system that is closed for inputs and outputs of energy and mass is called an isolated system, while a system that is closed to inputs and outputs of mass, but open to energy transfers, is denoted a closed system. A nonisolated system is a closed or open system. If an ecosystem was isolated, it would inevitably move toward thermodynamic equilibrium and become a dead system with no gradients to do work.

> This is the consequence of the second law of thermodynamics, which states that all systems will lose order and gain disorder, or expressed differently, all systems will gain entropy, lose energy that can do work to energy that cannot do work, i.e., all systems will lose exergy. As presented in Chapter 2, the thermodynamic equilibrium will correspond to $dG = 0$ and $dS = 0$ at a maximum S value and a minimum value of free energy.

The openness explains why an ecosystem can maintain life and stay far from thermodynamic equilibrium, because maintenance of life requires input of energy (or rather, free energy, i.e., energy that can do work), which of course only is possible if an ecosystem is at least nonisolated. As we have exchange of matter between the environment and the ecosystems, the ecosystems must be open. The ecological significance of the openness of ecosystems is the topic of Chapter 8, where more details about the ecological implications of the openness will be presented.

The use of the second law of thermodynamics for open systems is crucial, which is an important topic for this chapter. At first glance, it looks like ecosystems violate the first law of thermodynamics because they are moving away from thermodynamic equilibrium by formation of a biological structure, implying that they gain chemical energy. Ecosystems receive, however, energy as solar radiation, which delivers the energy for formation of the biological structure and also the energy needed for maintenance of the system far from thermodynamic equilibrium. Several proposals on how to apply the second law of thermodynamics on systems far from thermodynamic equilibrium have been given, as will be demonstrated in Section 4.5, but before we turn to this central issue for open systems, some basic implications of the openness of ecosystems will be discussed.

4.2 Physical Openness

An energy balance equation for ecosystems might be written as follows in accordance with the principle of energy conservation:

$$E_{cap} = Q_{evap} + Q_{resp} + \dots + \Delta E_{bio} \qquad (4.1)$$

Here E_{cap} is external energy captured per unit of time. A part of the incoming energy, solar radiation being the main source for the ecosystems on earth, is captured and a part is reflected unused, determining the albedo of the globe. See Figure 2.10 for an illustration of these relationships. The more biological structure an ecosystem possesses, the more of the incoming energy it is able to capture, i.e., the lower the albedo. The structure functions as an umbrella capturing the incoming solar radiation.

In ecosystem steady states, the formation of biological compounds (anabolism) is in approximate balance with their decomposition (catabolism). That is, in energy terms:

$$\Delta E_{bio} \approx 0, \text{ and } E_{cap} \approx Q_{evap} + Q_{resp} + \dots \tag{4.2}$$

The energy captured can in principle be any form of energy (electromagnetic, electrical, magnetic, chemical, kinetic, etc.), but for the ecosystems on earth the shortwave energy of solar radiation (electromagnetic energy) plays the major role. The energy captured per unit of time is, however, according to Equation (4.2) used to pay the cost of maintenance per unit of time $= Q_{evap} + Q_{resp} \dots$. The overall results of these processes require that E_{cap} be >0, which entails openness (or at least nonisolation). The following reaction chain summarizes the consequences of energy openness (Jørgensen et al., 1999): *source*: solar radiation → *anabolism* (charge phase): incorporation of high-quality energy, with entrained work capacity (and information), into complex biomolecular structures, entailing antientropic system movement away from equilibrium → *catabolism* (discharge phase): deterioration of structure involving release of chemical bond energy and its degradation to lower states of usefulness for work (heat) → *sink*: dissipation of degraded (low work capacity and high entropy) energy as heat to the environment (and, from earth, to deep space), involving entropy generation and return toward thermodynamic equilibrium.

The same chain can also be expressed in terms of matter: *source*: geochemical substrates relatively close to thermodynamic equilibrium → *anabolism*: inorganic chemicals are molded into complex organic molecules (with low probability, this means that the equilibrium constant for the formation process is very low, low entropy, and high distance from thermodynamic equilibrium) → *catabolism*: synthesized organic matter is ultimately decomposed into simple inorganic molecules again; the distance from thermodynamic equilibrium decreases, and entropy increases → *cycling*: the inorganic molecules, returned to near-equilibrium states, become available in the nearly closed material ecosphere of earth for repetition of the matter charge-discharge cycle.

Energy inflows to ecosystems are of high-quality energy with high contents of work and information and low entropy and are able to raise the organizational states of matter far from equilibrium. The outflows are, in contrast, sinks for energy and matter and have lower work capacity, are higher in entropy, and closer to equilibrium.

4.3 Ontic Openness

One of the key questions in natural science of the twentieth century was: Is the world deterministic—in the sense that if we would know the initial conditions in all details, could we also predict in all details how a system would develop—or is the world ontic open?

We cannot and will probably never be able to answer these two questions, but the world is under all circumstances too complex to enable us to *determine* in all details the initial conditions. The uncertainty relations similar to Heisenberg's uncertainty relations in quantum mechanics are also valid in ecology. This idea has been presented in Jørgensen (1988, 2002; Jørgensen et al., 2007), but will be summarized below, because the discussion in the next chapters is dependent on this uncertainty in our description of nature. The world may be ontic open = nondeterministic, because the universe has been created that way, or it may be ontic open = nondeterministic, because nature is too complex to allow us to know a reasonable fraction of the initial conditions even for a subsystem of an ecosystem. We shall most probably never be able to determine which of the two possibilities will prevail, but it is not of importance, because we have to accept ontic openness in our description of nature.

If we take two components and want to know all the relations between them, we would need at least three observations to show whether the relations were linear or non-linear. Correspondingly, the relations among three components will require 3*3 observations for the shape of the plane. If we have 18 components we would correspondingly need 3^{17}, or approximately 10^8, observations, which is a realistic but high number of observations. At present this is probably an approximate, practical upper limit to the number of observations that can be invested in one project aimed for one ecosystem. If we estimate the accuracy obtainable for one ecological observation to be 0.1 relatively, that is, as 10% standard deviation, we can in that case use all the observations to get one piece of information with a very high accuracy, obtaining an accuracy of $0.1/\sqrt{10^8} = 10^{-5}$.

This could be used to formulate a practical uncertainty relation in ecology; see also Jørgensen (1988):

$$10^5 * \Delta x / \sqrt{3^{n-1}} = 1 \tag{4.3}$$

where Δx is the relative accuracy of one relation, and n is the number of components examined or included in the model. If $n = 1$ in Equation (4.3), we use all observations to get one piece of very accurate information, $\Delta x = 10^{-5}$. If $n = 18$, $\Delta x = 0.1$, because there is only applied one observation per relationship, and with 18 components there are 3^{17} relationships, as discussed above.

If $n = 13$, for instance, $\Delta x \approx 0.01$, because we apply 100 repetitions per ecological observation and thereby get $\sqrt{100}$ times less uncertainty, or $0.1/10 = 0.01$, because the accuracy is inversely proportional to the square root of the number of observations.

Costanza and Sklar (1985) talk about the choice between the two extremes: knowing everything about nothing, or nothing about everything. The first refers to the use of all the observations on one relation to obtain a high accuracy and certainty, while the latter refers to the use of all observations on as many relations as possible in an ecosystem.

Equation (4.3) formulates a practical uncertainty relation, but of course, the possibility that the practical number of observations may be increased in the future cannot be excluded. Ever more automatic analytical equipment is emerging on the market. This means that the number of observations that can be invested in one project may be one, two, three, or even several magnitudes larger in one or more decades.

Example 4.1

For a project, we can apply $2 million for observations and one observation costs $0.1. An accuracy of 0.05 is needed for the development of a model and the usual standard deviation of 10% for one observation can be presumed. How many state variables can the model include?

Solution

In accordance with the project budget, we can make 20 million observations. We need four observations per relationship to obtain 5% accuracy. This means that we have to change the constant 10^5 in Equation (3.4) to $5{,}000{,}000^{0.5}/0.1 = 2.25*10^4$, corrresponding to the application of 5 million observations that are repeated four times but each gives an accuracy of 10% relatively. Notice that Equation (4.3) presumes the application of 100 million observations.

This gives the following equation for determination of the number of state variables, n:

$$2.25*10^4*0.1 = (3^{n-1})^{0.5}$$

or

$$3 + \log 2.25 = 0.5*(n - 1)*\log 3$$

or

$$n - 1 = 13.4$$

or 14 state variables can be selected for the model.

However, a theoretical uncertainty relation can be developed. If we go to the limits given by quantum mechanics, the number of variables will still be low compared to the number of components in an ecosystem.

One of Heisenberg's uncertainty relations is formulated as follows:

$$\Delta s*\Delta p \geq h/2\pi \tag{4.4}$$

where Δs is the uncertainty in determination of the place, and Δp is the uncertainty of the momentum. According to this relation, Δx of Equation (4.3) should be in the order of 10^{-17} if Δs and Δp are about the same. Another of Heisenberg's uncertainty relations may now be used to give the upper limit of the number of observations:

$$\Delta t*\Delta E \geq h/2\pi \tag{4.5}$$

where Δt is the uncertainty in time and ΔE in energy.

If we use all the energy that the earth has received during its lifetime of 4.5 billion years we get:

$$173*10^{15}*4.5*10^9*365.3*24*3{,}600 = 2.5*10^{34} \text{ J} \tag{4.6}$$

where $173*1{,}015$ W is the energy flow of solar radiation. Δt would therefore be in the order of 1^{-69} s. Consequently, an observation will take 10^{-69} s, even if we use all the energy that has been available on earth as ΔE, which must be considered the most extreme case.

The hypothetical number of observations possible during the lifetime of the earth would therefore be

$$4.5*10^9*365.3*3{,}600/10^{-69} \simeq \text{ of } 10^{85} \tag{4.7}$$

This implies that we can replace 10^5 in Equation (4.3) with 1,060 since $1^{-17}/\sqrt{10^{85}} = \sim10^{-60}$. If we use $\Delta x = 1$ in Equation (4.3) we get:

$$\sqrt{3}n^{-1} \leq 10^{60} \qquad\qquad (4.8)$$

or

$$n \leq 253$$

From these very theoretical considerations, we can clearly conclude that we shall never be able to obtain a sufficient number of observations to describe even one ecosystem in all detail. These results are completely in harmony with the Niels Bohr's complementarity theory. He expressed it as follows: It is not possible to make one unambiguous picture (model or map) of reality, as uncertainty limits our knowledge. The uncertainty in nuclear physics is caused by the inevitable influence of the observer on the nuclear particles, or rather, we do not know whether an electron is a particle or a wave or both at the same time, because an electron is so small that it cannot be described by use of our normal physical concepts, and it cannot be observed by the usual applied methods. In ecology the uncertainty is caused by the enormous complexity and variability, which will always make it impossible to perform the observations needed to make a complete description. Therefore, no map of reality is completely correct. There are many maps (models) of the same ecosystem, and the various maps or models reflect different viewpoints and different sets of observations, for instance, distribution of plants and soil properties, and so on. Accordingly, one model (map) does not give all the information and far from all the details of an ecosystem. In other words, the theory of complementarity is also valid in ecology.

The use of maps in geography is a good parallel to the use of models in ecology (Jørgensen and Bendoricchio, 2001). As we have road maps, aeroplane maps, geological maps, and maps in different scales for different purposes, we have in ecology many models of the same ecosystems, and we need them all if we want to get a comprehensive view of ecosystems. A map can furthermore not give a complete picture. We can always do the scale larger and larger and include more details, but we cannot get all the details, for instance, where all the cars of an area are situated just now, and if we could get the picture, a few seconds later it would be invalid because we want to map too many dynamic details at the same time. An ecosystem also consists of too many dynamic components to enable us to model all the components simultaneously, and even if we could, the model would be invalid a few seconds later, where the dynamics of the system has changed the picture.

In nuclear physics we need to use many different pictures of the same phenomena to be able to describe our observations. We say that we need a pluralistic view to cover our observations completely. Our observations of light, for instance, require that we consider light as waves as well as particles. The situation in ecology is similar. Because of the immense complexity we need a pluralistic view to cover a description of the ecosystems according to our observations. We need many models covering the different possible viewpoints, based on different sets of observations. We adopt what we could call model-dependent realisms (see Hawking, 2001). Scientific theory or our picture of the world is a model—generally of a mathematical nature. It includes rules that connect the elements of the model to observations. According to the model-dependent realism, it is pointless to ask whether a model is real, only whether it agrees with the observations. If two models agree with the observations, neither one can be considered more real than the other. A person can use whichever model is more convenient in the situation under consideration (see Hawking, 2001).

In addition to physical openness, there is also an epistemological openness inherent in the formal lenses through which humans view reality. Gödel's theorem, which was published in January 1931, introduces an epistemic openness in a very strong way. The theorem requires that mathematical and logical systems (i.e., purely epistemic, as opposed to ontic) cannot be shown to be self-consistent within their own frameworks, but only from outside. A logical system cannot itself (from inside) decide on whether it is false or true. This requires an observer from outside the system, and this means that even epistemic systems must be open.

We can distinguish between ordered and random systems. Many ordered systems have emergent properties defined as properties that a system possesses in addition to the sum of properties of the components—the system is more than the sum of its components. Wolfram (1984a, 1984b) calls these *irreducible systems* because their properties cannot be revealed by a reduction to some observations of the behavior of the components. It is necessary to observe the entire system to capture its behavior, because everything in the system is dependent on everything else due to direct and indirect linkages. The presence of irreducible systems is consistent with Gödel's theorem, according to which it will never be possible to give a detailed, comprehensive, complete. and comprehensible description of the world. Most natural systems are irreducible, which places profound restrictions on the inherent reductionism of science.

In accordance with Gödel's theorem, properties of order and emergence cannot be observed and acknowledged from within the system, but only by an outside observer. It is consistent with the proverb "You cannot see the forest through the trees," meaning that if you only see the trees as independent details inside the forest, you are unable to observe the system, a forest as a cooperative unit of trees. This implies that the natural sciences, aiming toward a description or ordering of the systems of nature, have meaning only for open systems. A scientific description of an isolated system, i.e., the presentation of an algorithm describing the observed, ordering principles valid for an isolated system, is impossible. In addition, sooner or later an isolated ontic system will reach thermodynamic equilibrium, implying that there are no ordering principles, but only randomness—entropy has reached its maximum and there is no longer exergy and eco-exergy left in the system, because it is homogenous and without gradients. We can infer from this that an isolated epistemic system will always ultimately collapse inward on itself if it is not opened to cross-fertilization from outside. Thomas Kuhn's account of the structure of scientific revolutions would seem to proceed from such an epistemological analogy of the second law.

This does imply (Jørgensen et al., 1999) that we cannot describe all open systems in all details. The only complete detailed and consistent description of a system is the system itself. We can furthermore never know if a system or subsystem is ordered or random, because we may not have found the algorithm describing the order. We can never know if it exists, or if we can find it later by additional effort. The challenge of modeling and model making is to find a simplified description of reality to be applied in a well-defined context, which is in accordance with our definition of life (Patten et al., 1997). In this chapter it is claimed that a useful definition of life is that living organisms are able to make models. A model is always a simplified or homomorphic description of some features of a system, but no model can give a complete or isomorphic description. Therefore, one might conclude that an infinite number of different models is required to realize a complete, detailed, comprehensive, and consistent description of any entire system. In addition, it is also not possible to compute or totally explain our thoughts and conceptions of our limited but useful description of open natural systems. Our perception of nature goes beyond what can be explained and computed. Nevertheless, it is possible for us to conceive irreducible (open) systems, though we cannot explain all the details of the system. This explains the applicability and usefulness of models in the adaptations of living things ("subjects"; Patten et al., 1997) to their environment. It also underlines that the models in the best case only will be able to cover one or a few out of many views of considered systems. If we apply the definition of life proposed in Patten et al. (1997), then life is things that make models—this implies that all organisms and species must make their way in the world based on only partial representations, limited by the perceptual and cognitive apparatus of each, and the special epistemologies or models that arise therefrom. The models are always incomplete, but sufficient to guarantee survival and continuance, or else extinction is the price of a failed model.

Following from Gödel's theorem, a scientific description can only be given from outside open systems. Natural science cannot be applied to isolated systems (the universe is considered open due to the expansion) at all. A complete, detailed, comprehensive, and consistent description of an open system can never be obtained. Only a partial though useful description (model) covering one or a few out of many views can be achieved.

Due to the enormous complexity of ecosystems we cannot, as already stressed, know all the details of ecosystems. When we cannot know all the details, we are not able to describe fully the initial stage and the processes that determine the development of the ecosystems—as expressed above, ecosystems are therefore irreducible. Ecosystems are not deterministic because we cannot provide all the observations that are needed to give a full determinsitic description. Or as expressed by Tiezzi (2003, 2006), ecosystems do play dice. This implies that our description of ecosystem developments must be open to a wide spectrum of possibilities. It is consistent with the application of chaos and catastrophe theory; see, for instance, Jørgensen (1992a, 1992b, 1994a, 1994b, 2002). Ulanowicz (1986, 1997) makes a major issue of the necessity for systems to be causally open in order to be living—the open possibilities may create new pathways for development that may be crucial for survival and further evolution in a nondeterministic world. He goes so far as to contend that a mature insight into the evolutionary process is impossible without a revision of our contemporary notions on causality. Ulanowicz and Abarca-Arenas (1997) and Ulanowicz et al. (2006) use the concept of propensity to come around the problem of causality. On the one side, we are able to relate the development with the changing internal and external factors of ecosystems. On the other side, due to the uncertainty in our predictions of development caused by our lack of knowledge about all details, we are not able to give deterministic descriptions of the development, but we can only indicate which propensities and development directions will be governing.

To conclude, ecosystems are ontic open. They are irreducible and due to their enormous complexity, which prohibits us to know all details, we will only be able to indicate the propensities and directions of their development. Ecosystems are not deterministic systems. Nevertheless, the more we know about ecosystems, the better we can describe their possible development as a consequence of different impacts on the ecosystems.

4.4 The Second Law of Thermodynamics Interpreted for Ecosystems

Let us first expand on the conclusions that we already have made to give more detail on the difference between isolated and open systems, and thereby understand better the application of the second law of thermodynamics on open systems such as ecosystems (Jørgensen et al., 1999).

If ecosystems were isolated, no energy or matter could be exchanged across their boundaries. The systems would spontaneously degrade their initially contained exergy (loss of work energy to heat energy at the temperature of the environment), increase their entropy, corresponding to a loss of order and organization, and increase in the randomness of their constituents and microstates. This dissipation process would cease at equilibrium, where no further motion or change would be possible. The physical manifestation would ultimately be a meltdown to the proverbial "inorganic soup" containing degradation products dispersed equi-probably through the entire volume of the system. All gradients of all kinds would be eliminated, and the system would be frozen in time in a stable, fixed configuration. The high-energy chemical compounds of biological systems, faced suddenly with isolation, would decompose spontaneously to compounds with high-entropy contents. The process would be progressive, to higher and higher entropy, and would by presence of oxygen end with a mixture of inorganic residues at high oxidation state—carbon dioxide and water, nitrates, phosphates, sulfates, etc. These simpler compounds could only, with an extremely low probability, be reconfigured into the complex molecules necessary to carry on life processes without the input of new low-entropy energy to be employed in biosynthesis. An isolated ecosystem could therefore, in the best case, sustain life for only a limited period of time, less than that required from the onset of isolation to reach thermodynamic equilibrium. This local situation is comparable to the "heat death" of the universe, seen by physicists of a century ago as the ultimate outcome of the second law of thermodynamics. Today, we are questioning the heat death due to the expanding universe (see Jørgensen, 2008). However, thermodynamic equilibrium is the global attractor for all physical processes isolated from their surroundings. Having reached it, no further changes are possible. In this "frozen" state even time would have no meaning, as its passage could not be verified by reference to any changes. Observations of properties could not be made, only inferred, because observation requires some kind of exchanges between the system and an observer. There would be no internal processes, because no gradients would exist to enable them. There would only be uninterrupted and uninterruptable stillness and sameness, which would never change. The system would be completely static at thermodynamic equilibrium. Thus, in a peculiar way, isolated systems can only be pure abstractions in reality, submitting neither to time passage, change, nor actual observation. They are the first "black holes" of physics, and the antithesis of our systems plus their environments, which are the core model for systems ecology. No ecosystem could ever exist and be known to us as an isolated system.

The change in entropy for an *open* system, dS_{system}, consists of an external, exogenous contribution from the environment, $d_e S = S_{in} - S_{out}$, and an internal, endogenous contribution due to system state, $d_i S$, which must always be positive by the second law of thermodynamics (Prigogine, 1980). Prigogine (1980) uses the concept of entropy and the second law of thermodynamics far from thermodynamic equilibrium, which is outside the framework of classical thermodynamics, but he uses the concepts only locally. Tiezzi (2003, 2006) uses entropy opposite what is the case in classical thermodynamics, not as a state variable. Use of entropy as a nonstate variable makes it feasible to use the concept for systems far from thermodynamic equilibrium.

There are three possibilities for the entropy balance, according to Prigogine and Tiezzi:

$$dS_{system}/dt = d_e S/dt + d_i S/dt > 0 \qquad (4.9)$$

$$dS_{system}/dt = d_e S/dt + d_i S/dt < 0 \qquad (4.10)$$

$$dS_{system}/dt = 0 \qquad (4.11)$$

The system loses order in the first case. Gaining order (case 2) is *only* possible if $-d_e S > d_i S > 0$. This means that if order is to be created in a system ($dS_{system} < 0$), $d_e S$ must be < 0, and therefore $S_{in} < S_{out}$.

Creation of order in a system must be associated with a greater flux of entropy out of the system than into the system. This implies that the system must be open or at least nonisolated.

Case 3, Equation (4.11), corresponds to a stationary situation, for which Ebeling et al. (1990) use the following two equations for the energy (U) balance and the entropy (S) balance:

$$dU/dt = 0 \text{ or } d_e U/dt = -d_i U/dt = 0 \qquad (4.12)$$

and

$$dS_{system}/dt = 0 \text{ or } d_e S/dt = -d_i S/dt = 0 \qquad (4.13)$$

Usually the thermodynamic processes are isotherm and isobar. This implies that we can interpret the third case (Equations (4.11)–(4.13)) by use of the free energy:

$$d_e G/dt = T \, d_i S/dt > 0$$

It means that a status quo situation for an ecosystem requires input of free energy to compensate for the loss of free energy and corresponding formation of heat due to maintenance processes, i.e., respiration and evapotranspiration. If the system is not receiving a sufficient amount of free energy, the entropy will increase. If the entropy of the system continues to increase, the system will approach thermodynamic equilibrium—the system will die. This is in accordance with Ostwald (1931):

Life without the input of free energy or energy that can do work is not possible.

The entropy produced by the life processes can be exported by three processes: (1) transfer of heat to the environment, (2) exchange of material with the environment, and (3) biochemical processes in the system. The first process (heat transfer) is of particular importance.

An energy flow of about 10^{17} W by solar radiation ensures the maintenance of life on earth. The surface temperature of the sun is 5,800 K and of the earth, on average, about 280 K. This implies that the following export of entropy per unit of time takes place from the earth to open space:

$$10^{17} \text{ W } (1/5,800 \text{ K} - 1/280 \text{ K}) \approx 4.10^{14} \text{ W/K} \qquad (4.14)$$

corresponding to 1 W/m^2 K.

Ecosystems can maintain a certain concentration of low-entropy compounds against the second-law dissipation gradient, because they are not isolated. Ecosystems receive a continuous supply of free energy or negentropy (potential entropy, not yet released [see Schrödinger, 1944]) from outside to compensate for the positive entropy produced internally as a consequence of the second law of thermodynamics ($d_iS > 0$). On earth, solar radiation is the main source of this input of free energy, negentropy, or low-entropy energy. The incoming energy has low entropy, while the outgoing energy has higher entropy.

All ordered structures require low entropy for maintenance, and therefore for a system to maintain structure or increase its internal order it must receive input of low-entropy energy from external sources. *Structure*, in this context, is a spatial or temporal order describable in terms of information theory; see also Chapter 6. Prigogine uses the term *dissipative structure* to denote self-organizing systems, thereby indicating that such systems dissipate energy (produce entropy) for the maintenance of their organization (order). The following conclusions are appropriate:

All systems, because they are subject to the second law of thermodynamics, are inherently dissipative structures. To offset the dissipative processes they require inputs of low-entropy energy to maintain or produce a more internally organized structure, measurable in terms of information content. Thus, all real systems must be open, or at least nonisolated.

Ecosystems, in common with all real systems, have, as previously noted, a global attractor state, thermodynamic equilibrium (see Jørgensen et al., 2000). Through their openness they avoid reaching this state by importing low entropy, or matter carrying information from their surroundings. This anabolism combats and compensates for the catabolic deterioration of structure; the two processes operate against each other. Note that the equilibrium attractor represents a resting or refractory state, one that is passively devolved to if system openness or nonisolation is compromised (Jørgensen et al., 1999). The term is also commonly used to express the situation when a system is actively pushed or "forced" toward a *steady state*. Though widespread, we do not subscribe to this usage and make a distinction between steady states and equilibria.

As observed above, a steady state is a forced (nonzero input) condition; there is nothing "attractive" about it. Without a proper forcing function it will never be reached or maintained. A steady state that is constant may appear an equilibrium, but it is really far from equilibrium and maintained by a steady input of energy or matter. We regard equilibrium as a zero-input condition. What are often recognized as local attractors in mathematical models really have no counterparts in nature. Steady states are forced conditions, not to be confused with unforced equilibria, which represent states to which systems settle when they are devoid of inputs. The only true natural attractor in reality, and it is global, is the unforced thermodynamic equilibrium.

4.5 The Third Law of Thermodynamics Applied on Open Systems

The first law of thermodynamics is often applied to ecosystems, first of all when energy balances of ecosystems are made. Also, the second law of thermodynamics is applied to ecosystems when we consider the entropy production of ecosystems as a consequence of the maintenance of the system far from thermodynamic equilibrium. This section is concerned with the application of the third law of thermodynamics on ecosystems, which has a more indirect or theoretical application.

The lesser known third law of thermodynamics states that the entropies, S^0, of pure chemical compounds are zero, and that entropy production, ΔS^0, by chemical reactions between pure crystalline compounds is zero at absolute temperature, 0 K. Remember in this context that temperature is a measure of how fast atoms are moving, where 0 K corresponds to no movement at all. The third law implies, since both $S^0 = 0$ (absolute order) and $\Delta S^0 = 0$ (no disorder generation), that disorder does not exist and cannot be created at absolute zero temperature.

But at temperatures of >0 K disorder can exist ($S_{system} > 0$) and be generated ($\Delta S_{system} > 0$). The third law defines the relation between entropy production, ΔS_{system}, and the Kelvin temperature, T:

$$\Delta S_{system} = \int_0^T \Delta c_p \, d\ln T + \Delta So \qquad (4.15)$$

where Δc_p is the increase in heat capacity by the chemical reaction. Since order is absolute at absolute zero, its further creation is precluded there.

At higher temperatures, however, order can be created and maintained. It has been stated a few times that it is *necessary* for an ecosystem to transfer the generated heat (entropy) to the environment and to receive low-entropy energy (solar radiation) from the environment for formation of a dissipative structure. The next obvious question would be: Will an energy source and sink also be *sufficient* to initiate formation of a dissipative structure, which can be used as a source for entropy combating processes?

The answer is yes. It can be shown by the use of simple model systems and basic thermodynamics (see Morowitz, 1968).

He shows that a flow of energy from sources to sinks leads to an internal organization of the system and to the establishment of element cycles. The type of organization is, of course, dependent on a number of factors: the temperature, the elements present, the initial conditions of the system, and the time available for the development of the organization. It is characteristic for the system, as pointed out above, that the steady state of an open system does *not* involve chemical equilibrium.

An interesting illustration of the creation of an organization (dissipative structure) as a result of an energy flow through ecosystems concerns the possibilities to form organic matter from the inorganic components that were present in the primeval atmosphere. Since 1897 many simulation experiments have been performed to explain how the first organic matter was formed on earth from inorganic matter. All of them point to the conclusion that energy interacts with a mixture of gases to form a large set of randomly synthesized organic compounds. Most interesting is perhaps the experiment performed by Stanley Miller and Harold Urey at the University of Chicago in 1953, because it showed that amino acids can be formed by sparking a mixture of CH_4, H_2O, NH_3, and H, corresponding approximately to the composition of the primeval atmosphere (see Figure 4.1).

Prigogine and his colleagues have shown that open systems that are exposed to an energy throughflow exhibit coherent self-organization behavior and are known as dissipative structures. Formations of complex organic compounds from inorganic matter,

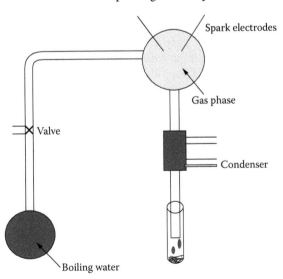

FIGURE 4.1 (See color insert). The apparatus used by Stanley Miller and Harold Urey to simulate reactions in the primeval atmosphere.

as mentioned above, are typical examples of self-organization. Such systems can remain in their organized state by exporting entropy outside the system, but are dependent on outside free energy fluxes to maintain their organization, as was already mentioned and emphasized above. The organization is inexorably torn down according to the second law of thermodynamics. Only a flow of free energy can compensate for this decomposition of the organization. Glansdorff and Prigogine (1971) have shown that the thermodynamic relationship of far from equilibrium dissipative structures is best represented by coupled nonlinear relationships, i.e., autocatalytic positive feedback cycles.

Given this necessary condition, simple energy flow through a system would provide a sufficient condition, and creation of order is inevitable. On earth, the surface temperature difference between sun and our planet guarantees this. Morowitz (1968) showed, as mentioned above, that energy throughflow is sufficient to produce cycling, a prerequisite for the ordering processes characteristic of living systems.

A system at 0 K, on the other hand, is without any creative potential, because no dissipation of energy can take place at this temperature. A temperature greater than 2.726 ± 0.01 K, where 2.726 K is the temperature of deep space, is therefore required before order can be created. At 0 K the world is dead and still, as the temperature is a measure of the velocity of atoms. The so-called Bose-Einstein condensate is formed, where all atoms are the same and behave like one single atom. This was predicted by Bose and Einstein in the 1920s, but has recently been shown experimentally at temperatures very close to 0 K.

The velocity of atoms is zero at 0 K by definition, and therefore determined without uncertainty. This explains that the position is undetermined according to Heisenberg's uncertainty equation (4.4). At 0 K there is therefore no structure, no gradients, and no complexity. No entropy can be formed because all mass is everywhere and nowhere and without form and structure. There is no disorder to create, and therefore no entropy to produce. The system is trapped between complete order because all the mass occupies all the space, and complete disorder because all the space is occupied by mass—a complete dissipation has taken place. At 0 K, no creativity is possible, no differences (gradients), no structure, and even no physical activity, because all velocities are zero. Everything is dull and dead. Even the light has stopped. Time has no meaning, because time is determined by the rate of changes and there are no changes.

These extreme conditions at 0 K elucidate the meaning of the concept entropy. Entropy is on the one side the price we have to pay for order, structure, organization, and creativity, and without entropy, there would be no order, structure, organization, and creativity.

Moreover, it explains the meaning behind the second law of thermodynamics—because heat is an energy form that is generated by transformation of all other energy forms, and because a 100% effective transformation of heat to work cannot take place. It would require that we had a temperature difference from the temperature T to the temperature 0 K. It means that we would be able to have a reservoir at 0 K, which is impossible. So,

Energy that can do work is, according to the second law of thermodynamics, inexorably lost to heat energy at the temperature of the environment, which means that it cannot do work. This is the condition that is imposed on us and all ecosystems, or expressed differently, time (and all reactions) is irreversible. All processes imply loss of exergy or work capacity.

4.6 Dissipative Structure and Eco-Exergy

As an ecosystem is nonisolated, the entropy changes during a time interval, dt, can be decomposed into the entropy flux due to exchanges with the environment and the entropy production due to the irreversible processes inside the system, such as diffusion, heat conduction, and chemical reactions; see Section 3.4. The entropy equation (4.9) can be expressed by use of exergy:

$$Ex/dt = d_e Ex/dt + d_i Ex/dt \qquad (4.16)$$

where $d_e Ex/dt$ represents the exergy input to the system and $d_i Ex/dt$ is the exergy consumed (is negative) by the system for maintenance, etc.

Equation (4.16) shows, among other things, that systems can only maintain a non-equilibrium steady state by compensating the internal exergy consumption with a positive exergy influx ($d_e Ex/dt > 0$). If $d_e Ex > -d_i Ex$ (the exergy consumption in the system), the system has surplus exergy input, which may be utilized to construct further order in the system, or as Prigogine (1980) calls it: dissipative structure. The system will thereby move further away from the thermodynamic equilibrium. The evolution shows that this situation has been valid for the ecosphere on a long-term basis. In spring and summer ecosystems are in the typical situation that $d_e Ex$ exceeds $-d_i Ex$. If $d_e Ex < -d_i Ex$, the system cannot maintain the order already achieved, but will move closer to the thermodynamic equilibrium; i.e., it will lose order. This is the situation for ecosystems during fall and winter in temperate climates or due to environmental disturbances.

It is therefore of interest to set up an exergy balance in addition to an energy balance. Loss of exergy needs our focus because it means that energy that can do work is lost and replaced by energy that cannot do work but is heat lost to the environment at the temperature of the environment. The latter form of energy may also be called anergy (Cerbe and Hoffmann, 1996). Energy is the sum of exergy and anergy: energy = exergy + anergy, and in accordance with the second law of thermodynamics, anergy is always positive for all processes.

4.7 How to Calculate Exergy of Organic Matter and Organisms

The following expressions for the exergy per unit of volume have been presented; see Equation (3.15):

$$Ex = RT \sum_{i=0}^{i=n} c_i \ln c_i/c_{io} \qquad [ML^{-1}T^{-2}] \qquad (4.17)$$

where R is the gas constant, T is the temperature of the environment, and c_i is the concentration of the i-th component expressed in a suitable unit; e.g., for phytoplankton in a lake c_i could be expressed as mg/L or as mg/L of a focal nutrient. c_{io} is the concentration of the i-th component at thermodynamic equilibrium and n is the number of components.

c_{io} is very low for living components because the probability that living components are formed at thermodynamic equilibrium is very low. This implies that living components get a high exergy. c_{io} is not zero for organisms, but will correspond to a very low probability of forming complex organic compounds spontaneously in an inorganic soup at thermodynamic equilibrium. c_{io}, on the other hand, is high for inorganic components, and although c_{io} is still low for detritus, it is much higher than for living components.

Shieh and Fan (1982) have suggested to estimate the exergy of structurally complicated material on the basis of the elementary composition. This, however, has the disadvantage that a higher organism and a microorganism with the same elementary composition will get the same exergy, which is in complete disagreement with the lower probability to form a more complex organism, i.e., the lower concentration of c_{io} in Equation (4.18).

The problem related to the assessment of c_{io} has been discussed and a possible solution proposed in Jørgensen et al. (1995). For dead organic matter, detritus, which is given the index 1, can be found from classical thermodynamics to be approximately 18.7 kJ/g on average.

For the biological components, 2, 3, 4, ..., N, the probability, p_{io}, consists at least of the probability of producing the organic matter (detritus), i.e., p_{1o}, and the probability, $p_{i,a}$, to find the correct composition of the enzymes determining the biochemical processes in the organisms. Living organisms use 20 different amino acids, and each gene determines on average the sequence of about 700 amino acids (Li and Grauer, 1991). $p_{i,a}$, can be found from the number of permutations among which the characteristic amino acid sequence for the considered organism has been selected. This means:

$$p_{i,a} = a^{-Ng_i} \quad [-] \qquad (4.18)$$

where a is the number of possible amino acids = 20, N is the number of amino acids determined by one gene = 700, and g_i is the number of nonnonsense genes. The following two equations are available to calculate P_i:

$$P_{io} = P_{1o} P_{i,a} = P_{1o} a^{-Ng} \approx p_{1o} \cdot 20^{-700} \quad [-] \qquad (4.19)$$

and the exergy contribution of the i-th component can be found by combining Equations (4.18) and (4.19):

$$Ex = RT \, c_i \ln c_i/(p_{1o} \, a^{-Ng} \, c_{oo}) = (\mu_1 - \mu_{1o}) \, c_i - c_i \ln p_{i,a} = (\mu_1 - \mu_{1o}) \, c_i - c_i \ln (a^{-Ng_i})$$

$$(4.20)$$

$$= 18.7 c_i + 700 \, (\ln 20) \, c_i \, g_i \, [ML^{-1}T^{-2}]$$

The total eco-exergy can be found by summing up the contributions originated from all components. The contribution by inorganic matter can be neglected as the contributions by detritus and, even to a higher extent, from the biological components are much higher due to an extremely low concentration of these components in the reference system. The contribution by detritus, dead organic matter, is 18.7 kJ/g times the concentration (in g/unit of volume), while the exergy of living organisms consists of

$$\text{Ex}_i^{\text{chem}} = 18.7 \text{ kJ/g times the concentration } c_i \text{ (g/unit of volume)}$$

and

$$\text{Ex}_i^{\text{bio}} = \text{RT} \,(700 \ln 20) \, c_i \, g_i = \text{RT} \, 2{,}100 \, g_i \, c_i \qquad (4.21)$$

$R = 8.34$ J/mole, and if we presume a molecular weight of, on average, 10^5 for the enzymes, we obtain the following equation for Ex_i^{bio} at 300 K:

$$\text{Ex}_i^{\text{bio}} = 0.0529 \, g_i \, c_i \qquad (4.22)$$

where the concentration now is expressed in g/unit of volume and the exergy in kJ/unit of volume.

For the entire system the exergy, Ex-total can be found as

$$\text{Ex-total} = 18.7 \sum_{i=1}^{N} c_i + 0.0529 \sum_{i=1}^{N} c_i \, g_i \; [\text{ML}^{-1}\text{T}^{-2}] \qquad (4.23)$$

where g for detritus ($i = 1$) of course is 0.

Table 4.1 gives an overview of the exergy of various organisms expressed by weighting factor β that is introduced to be able to cover the exergy for various organisms in the unit detritus equivalent or chemical exergy equivalent per unit of volume or unit of area (we find the exergy density):

$$\text{Ex} - \text{total-density} = \sum_{i=1}^{N} \beta_i c_i \; (\text{as detritus equivalents at the g temperature T} = 300 \text{ K} \qquad (4.24)$$

The β-value embodied in the biological/genetic information is found on the basis of Equations (4.20) and (4.22) by adding the chemical and biological contributions to calculate the eco-exergy. Detritus has in accordance with Equation (4.23) a β-value of 1.0. By multiplication of the result obtained by Equation (4.24) by 18.7, the exergy can be expressed in kJ per cubic meter or square meter, assuming that Equation (4.26) is giving the grams detritus equivalent.

The biological exergy is found (1) from the knowledge of the entire genome for the 11 organisms and (2) for a number of other organisms by use of a correlation between various complexity measures and the information content of the genome (see Jørgensen et al., 2005). The values of β for various organisms have been discussed in *Ecological Modelling* by several papers, but the latest published values in Jørgensen et al. (2005) probably come closest to the true β-values. The previous β-values were generally lower, but the relative values between two organisms have not been changed very much by the recently published β-values. As the β-values have been used consequently as relative measures, the previous results obtained by exergy calculations are therefore still valid. Weighting factors defined as the exergy content relative to detritus (see Table 4.1) may be considered quality factors reflecting how developed the various groups are and to what extent they contribute to the exergy due to their content of information, which is reflected in the computation. This is, completely according to Boltzmann (1905), what gave the following relationship for the work, W, that is embodied in the thermodynamic information:

TABLE 4.1   β-Values = Exergy Content Relative to the Exergy of Detritus;
β-Values = Eco-Exergy Content Relative to the Eco-Exergy of Detritus

Organisms	Plants	Animals	
Detritus	1.00		
Viroids	1.0004		
Virus	1.01		
Minimal cell	5.0		
Bacteria	8.5		
Archaea	13.8		
Protists (algae)	20		
Yeast	17.8		
		33	Mesozoa, placozoa
		39	Protozoa, amoeba
		43	Phasmida (stick insects)
Fungi, molds	61		
		76	Nemertina
		91	Cnidaria (corals, sea anemones, jelly fish)
Rhodophyta	92		
		97	Gastroticha
Prolifera, sponges	98		
		109	Brachiopoda
		120	Plathyhelminthes (flatworms)
		133	Nematoda (round worms)
		133	Annelida (leeches)
		143	Gnathostomulida
Mustard weed	143		
		165	Kinorhyncha
Seedless vascular plants	158		
		163	Rotifera (wheel animals)
		164	Entoprocta
Moss	174		
		167	Insecta (beetles, flies, bees, wasps, bugs, ants)
		191	Coleodiea (sea squirt)
		221	Lepidoptera (butterflies)
		232	Crustaceans
		246	Chordata
Rice	275		
Gymnosperms (incl. pinus)	314		
		310	Mollusca, bivalvia, gastropodea
		322	Mosquito
Flowering plants	393		
		499	Fish
		688	Amphibia
		833	Reptilia

TABLE 4.1 (Continued) β-Values = Exergy Content Relative to the Exergy of Detritus;
β-Values = Eco-Exergy Content Relative to the Eco-Exergy of Detritus

Organisms	Plants	Animals	
		980	Aves (birds)
		2,127	Mammalia
		2,138	Monkeys
		2,145	Anthropoid apes
		2,173	*Homo sapiens*

Source: Jørgensen, S.E., et al., *Ecol. Model.*, 185, 165–176, 2005.

$$W = RT \ln M \qquad (ML^2T^{-2}) \qquad (4.25)$$

where M is the number of possible states, among which the information has been selected. M is as seen for species the inverse of the probability to obtain spontaneously the amino acid sequence valid for the enzymes controlling the life processes of the considered organism.

Notice that the β-values are based on eco-exergy expressed in detritus equivalents per gram of biomass, because by multiplication by the biomass concentration per liter you will obtain the eco-exergy in detritus equivalents per liter. The β-values therefore express the specific eco-exergy, which according to Section 3.5 is equal to RTK, where K is Kullbach's measure of information. R is 8.34 J/mole K or 8.34 10^{-8} kJ/g presuming a molecular weight of 10^5. The chemical exergy is, of course, to be added. If we want to use the unit detritus equivalents instead of kJ, a division by 18.7 is needed.

This means that the β-values or the specific eco-exergy in detritus equivalents =

$$RTK = 8.34*300*10^{-8} \ln 20*AMS/18.7 = 4.00*10^{-6}*AMS \qquad (4.26)$$

where AMS is the number of amino acids in a coded sequence and ln 20 = 3.00.

The Kullback measure of information covers the gain in information, when the distribution is changed from p_{io} to p_i. Note that K is a specific measure (per unit of matter).

Viruses have coded about 2,500 amino acids. The β-value is therefore only 1.01. The smallest known agents of infectious disease are short strands of RNA. They can cause several plant diseases and are possibly implicated in enigmatic diseases of man and other animals. Viroids cannot encode enzymes. Their replication therefore relies entirely on enzyme systems of the host. Viroids typically have a nucleotid sequence of 360, which means that they would be able to code for 90 amino acids, although they are not translated. The β-value can therefore be calculated to be 1.0004. It can be discussed whether viroids should be considered living material.

Eco-exergy calculated by use of the above shown equations for ecosystems has some shortcomings:

1. We have made some minor approximations in the equations presented above.
2. We do not know the nonnonsense gene and the details of the entire genome for all organisms.

3. We calculate only in principle the exergy embodied in the proteins (enzymes), while there may be other components of importance for the life processes that should be included. These components, however, contribute less to the exergy than the enzymes, and the information embodied in the enzymes controls the formation of these other components, for instance, the hormones. It cannot be ignored that these components will contribute to the total exergy of the system.
4. We do not include the exergy of the ecological network when we calculate the eco-exergy for an ecosystem. If we calculate the exergy of models, the network will always be relatively simple and the contribution coming from the information content of the network is therefore minor, but the information content of a real ecological network may be significant due to its high complexity.
5. We will always use a simplification of the ecosystem, for instance, by a model or a diagram or similar. This implies that we only calculate the exergy contributions of the components included in the simplified image of the ecosystem. The real ecosystem will inevitably contain more components that are not included in our calculations.

It is therefore proposed that the exergy found by these calculations always should be considered a *relative minimum eco-exergy index* to indicate that there are other contributions to the total exergy of an ecosystem, although they may be of minor importance. In most cases, however, a relative index is sufficient to understand and compare the reactions of ecosystems, because the absolute exergy content is irrelevant for the reactions and cannot be determined due to the extremely high complexity. It is in most cases the change in eco-exergy that is of importance to understand the ecological reactions.

The weighting factors presented in Table 4.1 have been applied successfully in several structurally dynamic models, and furthermore in many illustrations of the maximum exergy principle, which will be presented in Chapters 6 and 7. The relatively good results in application of the weighting factors, in spite of the uncertainty of their more exact values, seem only to be explicable by the robustness of the application of the factors in modeling and other quantifications. The differences between the factors of the microorganism, vertebrates, and invertebrates are so clear that it seems not to be important whether the uncertainty of the factors is very high—the results are robust.

On the other hand, it would be important progress to get better weighting factors from a theoretical point of view, but also because it would enable us to model the competition between species that are closely related.

There is no doubt that the right estimation of β-values should be based on the number of proteins that are controlling the processes in the cells of various organisms. The knowledge to all human genomes is available today, and it has been found that the number of genes carrying nonrepetitive information is not 250,000, as previously applied, but rather about 40,000. On the other hand, it is also clear that the number of amino acids controlled by one gene is more than 700 for *Homo sapiens*. The total information in the genes is under all circumstances applied in the latest calculation in Table 4.1 of the β-values. The high information content can be explained by the number of amino acids per gene, which for some genes is as high as 38,000 (Hastie, 2001). The β-values in Table 4.1 can be considered the best estimation to date (2011). The primitive cell is indicated with a β-value of 5.0, while 5.8 was applied previously. Recent results have determined that the oceanic bacterium SAR 11 only has about 1,354 genes determining about 948,000 amino acids, which would correspond to a β-value of 4.88. It is therefore approximately correct to use 5.0 in Table 4.1.

The importance of the proteins can be seen from the intensive analytical work to find the composition of the human proteins, or as they are called now, the genetically determined proteins, proteoms, a word that is used to underline that the genetically produced proteins are of particular importance (Haugaard Nielsen, 2001). The great interest for the proteoms is due to their control of the life processes. Many proteoms may be the medicine of the future. The application of enzymes in industrial productions is just in its infancy, as there is enormous potential to control many more industrial processes by enzymes.

Example 4.2

The state variables of the P-cycle in Figure 2.8 have the following concentrations: Ortho P, 0.1 mg/L; Phyt-P, 0.1 mg/L; Zoopl-P, 0.02 mg/L; Fish-P, 0.0025 mg/L; Det-P, 0.025 mg/L; Exch. P, 0.022; and Por-P, 0.4 mg/L.

Calculate the eco-exergy in kJ/L of lake water, when it is presumed that P for phytoplankton, zooplankton, detritus, fish, and exchangeable P is 1%.

It is presumed that organic matter has 18.7 kJ/g free energy.

Solution

Equation (4.26) can be used directly as we know the transformities for the state variables.

(0.1*0 + 0.1*100*20 + 0.02*100*163 (Rotifera represents zooplankton) + 0.0025*100*499 + 0.025*100*1 + 0.022*100*1 + 0.4*0)*18.7 J/L = 12.252 kJ/L

The key to finding better β-values may be the proteoms. However, our knowledge about the number of proteoms in various organisms is very limited—more limited than for the number of information genes. It may be possible, however, to put together our knowledge about nonnonsense genes, the overall DNA content, the limited knowledge about the number of proteoms and the evolution tree, and see some pattern that could be used to give better but still very approximate β-values at this stage. For *Homo sapiens* it is presumed that 200,000 proteoms are produced by the cells and that they contain about 15,000 amino acids on average (Haugaard Nielsen, 2001). It would give a biological exergy that is $1 + 4.000 \times AMS \times 10^{-6}$, or 12,000, or considerably higher than the value in Table 4.1. The numbers used in these calculations are, however, still very uncertain. The other values in the table may, however, be changed similarly, which would not influence the results obtained by the use of eco-exergy, as it is the relative values that count in the application of eco-exergy for explanation of ecological observations and in structural dynamic modeling, as will be shown several times in the later chapters.

The free energy of oxidation, ΔG, in solution of the amino acid chains making up the enzymes for various living organisms has been found. The oxidation is to inorganic components, characterizing a reference system consisting of air, water, and sediment with the characteristic composition of the earth 4 billion years ago (see Ludovisi and Jørgensen, 2009). This quantity measures the free energy that an organism has invested in information and control of the life processes. This is because the oxidation (decomposition) of the amino acids and the peptide bonds of the enzymes produces the components that are characteristic for the reference system and for the components of a dead homogenous ecosystem without gradients at thermodynamic equilibrium.

Figure 4.2 shows the correlation between the β-value and the free energy invested by different organisms in the amino acid sequence of the enzymes. The correlation coefficient is >0.9999, which is not surprising because the β-value is calculated from the genomes, which determine the amino acid sequence. The strong correlation can, however, be considered a strong support for the statement that the β-value is expressing the information embodied in different organisms. Eco-exergy expresses the product of the biomass and the information for the organisms listed in Table 4.1. On the other hand, it would be beneficial to get more information about the genomes and the proteoms of various organisms to be able to obtain more reliable β-values and for a wider spectrum of organisms.

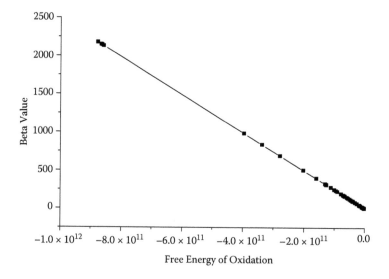

FIGURE 4.2 The found correlation between the free energy (kJ/mole) of oxidizing the amino acids that in the right sequence make up the enzymes controlling the life processes for various organisms and the β-values (see Table 4.1). By oxidation, an air, water, and sediment system of a composition close to the presumed equilibrium system on the earth 4 billion years ago represents a dead homogenous system, corresponding to the eco-exergy definition. (For further details, see Jørgensen et al., 2010.)

4.8 Why Have Living Systems Such a High Level of Exergy?

What is life? Most scientists would agree that life is:

1. The ability to metabolize, i.e., to draw nutrients from the environment and convert them from chemical energy into other forms of energy, useful biochemical compounds, and excrete waste products. The information embodied in the genes can be utilized to perform important life processes.
2. The ability to reproduce.

These two abilities are rooted in an enormous amount of (useful) information, that is able to control the processes needed to metabolize and reproduce. The β-values account for the information embodied in organisms.

A frog of 20 g will have an exergy content of $20 \times 18.7 \times 688$ kJ ≈ 257 GJ, while a dead frog will have an exergy content of only 374 kJ, although they have the same chemical composition, at least for a few seconds after the frog has died. The difference is rooted in the information, or rather the difference in the useful information. The dead frog has the information a few seconds after its death (the amino acid composition has not yet been decomposed), but the difference between a live frog and a dead frog is the ability to utilize the enormous information stored in the genes and the proteoms of the frog. So, the difference in eco-exergy between a live and a dead frog may define life as characterized by the ability to utilize the information in the amino acid sequence.

The amount of information stored in a frog is really surprisingly high. The number of amino acids placed in the right sequence is 84,000,000, and for each of these 84,000,000 amino acids the number of possibilities is 20. This amount of information in every cell, and there are trillion of cells, is able to ensure reproduction and is transferred from generation to generation, which ensures that evolution can continue because what is already a favorable combination of properties is conserved through the genes.

Because of the very high number of amino acids, 84,000,000, it is not surprising that there will always be a minor difference from frog to frog of the same species in the amino acid sequence. It may be a result of mutations or of a minor mistake in the copying process. This variation is important because it gives possibilities to "test" which amino acid sequence gives the best result with respect to survival and growth. This variability is a prerequisite for Darwin's theory. The best—representing the most favorable combination of properties—will have the highest probability of survival and give the most growth, and the corresponding genes will therefore prevail. Survival and growth mean more eco-exergy, and therefore a bigger distance from the thermodynamic equilibrium. Eco-exergy could therefore be used as a thermodynamic function, which could be used to quantify Darwin's theory. This is the idea behind the application of structurally dynamic models, which will be further discussed in Chapter 7.

4.9 Summary of Important Points in Chapter 4

1. Ecosystems are physically open systems. They receive an input of energy with a high content of exergy and a low content of entropy, which is applied to cover the maintenance of the ecosystems far from thermodynamic equilibrium. The energy is conserved in ecosystems, but the outflow of energy has a lower content of exergy (energy that can do work) and a higher content of entropy than the inflowing energy (mainly from solar radiation).

2. In accordance with the second law of thermodynamics, all processes in ecosystems will inevitably imply a loss of exergy to heat at the temperature of the environment, which means that energy that can do work is lost to energy that cannot do work. All processes therefore imply loss of exergy and production of entropy, and all processes are irreversible.

3. Ecosystems have an enormous complexity, which prohibits us from knowing all details. They are also irreducible and we are able to indicate the propensities and directions of their development. Ecosystems are nondeterministic systems. Nevertheless, the more we learn about ecosystems, the better we can describe the possible development as a consequence of different impacts on the ecosystems.

4. Eco-exergy density (per unit of volume or per unit of area) for ecosystems is found by

$$Ex - \text{total density} = \sum_{i=1}^{N} \beta_i c_i \text{ (as detritus equivalent)}$$

$$= 1 + 4.00 \times 10^{-6} * AMS$$

(where AMS is the number of amino acids in a coded sequence)

By multiplication by 18.7 the results in kJ/unit of volume or area can be found, because 1 g of detritus contains 18.7 kJ.

   The β-value expresses the free energy of the information stored in the amino acid sequence of the respective organisms.

Exercises/Problems

1. Give at least three different formulations of the second law of thermodynamics that could be useful in ecology.

2. It can be shown that power and eco-exergy are very well correlated. Explain why this correlation is probable and useful.

3. A wetland has per square meter 2,500 g of wild rice, 400 g of detritus, 0.2 g of crabs, 2 g of insects, 0.2 g of fish, and 0.1 g of frogs. Calculate the eco-exergy density of the wetland in MJ per square meter.

4. Explain why there is a very big difference between the eco-exergy of a dead frog and that of a living frog.

5. Explain why the maximum growth rate of an ecosystem in the northern hemisphere usually is recorded in late June, why the maximum biomass usually is recorded in early August, and the minimum biomass in early February. Explain these seasonal variations with reference to the second law of thermodynamics.

6. Eco-exergy can be applied as a holistic indicator for the ecosystem health (see Jørgensen et al., 2010). Give the reasonability of this application of eco-exergy.

7. What is the sign of entropy change for denaturation of proteins?

8. Two gas bulbs are separated by a stopcock, which is closed. In the bulb is 1 mole of nitrogen and the other bulb is empty. What is the change in entropy and free energy when the stopcock is opened? How much work could be performed?

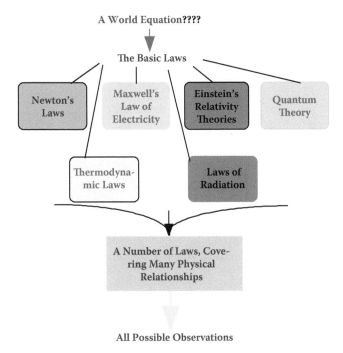

FIGURE 1.1 In physics, it is attempted to explain all possible observations by use of about 25 fundamentals laws and a number of other laws and rules that are deduced from the 25 basic laws. It is a dream in physics to find a world equation that can be used to explain the approximate 25 basic laws, but the dream has not yet been realized.

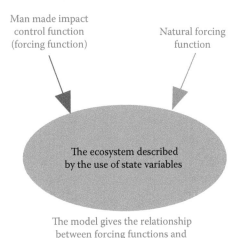

FIGURE 1.2 Ecological models give the relationships between forcing functions and state variables.

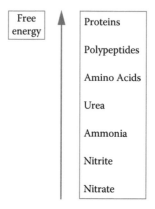

FIGURE 2.1   The levels of free energy for various biologically important nitrogen compounds.

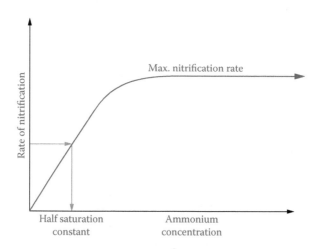

FIGURE 2.3   Graph of the Michaelis-Menten equation. In this case, the rate of nitrification vs. the concentration of the substrate, ammonium, is shown. The same graph is obtained for the growth rate of phytoplankton vs. the inorganic reactive phosphorus concentration, provided that phosphorus is the limiting nutrient.

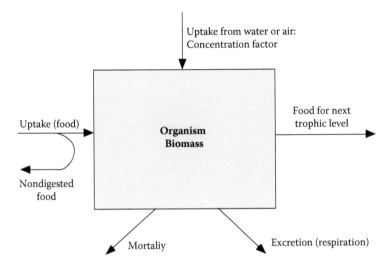

FIGURE 2.5 The mass balance principle applied on an organism. The inputs are food minus the nondigested food, and the outputs are mortality, excretion, and possible use of the food for the next level in the food chain.

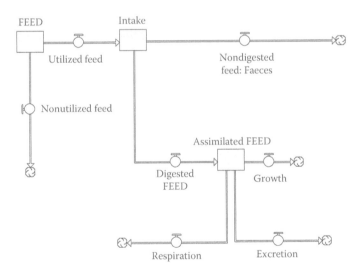

FIGURE 2.6 The available food may be either utilized or not utilized. The food eaten (intake) is either digested or not digested. The assimilated food is used for growth, respiration (maintenance of the structure far from thermodynamic equilibrium), or excreted. The illustration has adopted the symbols used by the software STELLA, where boxes are state variables and pipelines with valves are processes.

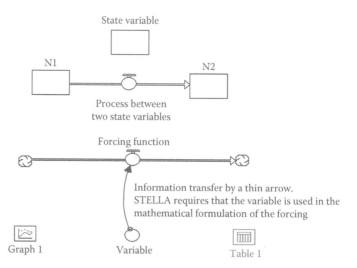

State variable

N1

N2

Process between
two state variables

Forcing function

Information transfer by a thin arrow.
STELLA requires that the variable is used in the
mathematical formulation of the forcing

Graph 1   Variable   Table 1

FIGURE 2.7 The components used for construction of a STELLA diagram: state variables, forcing functions, processes, information transfer, and auxiliary variables. It is possible to get the results as tables and graphs, as indicated in the figure. The diagram is useful to illustrate the dynamics of ecosystems.

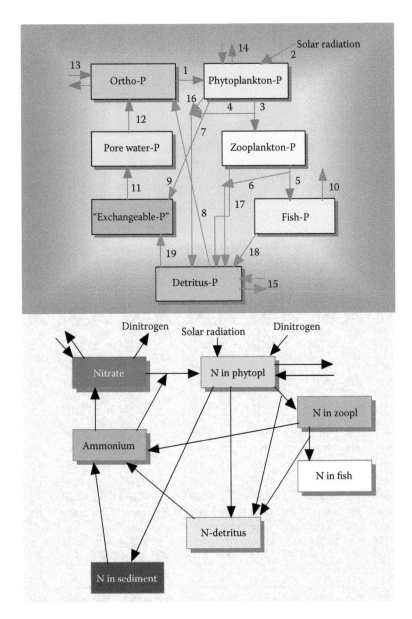

FIGURE 2.8 The phosphorus and nitrogen cycle in a lake, using a simple food chain (nutrient-phytoplankton-zooplankton-fish) to describe the processes of the cycle. The cycles are closed by mineralization of detritus and by release of nutrient by the sediment. (From Jørgensen, S.E., and Bendoricchio, G., *Fundamentals of Ecological Modelling*, 3rd ed., Elsevier, Amsterdam, 2001.)

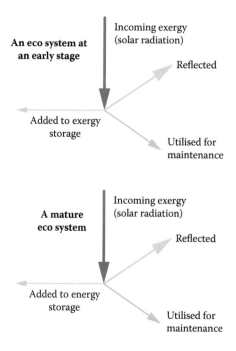

FIGURE 2.10 The utilization of energy is compared for the early and mature stages. The width of the arrows gives the approximate relative amount of energy. The reflected sunlight is less for the mature stage, but also less of the incoming solar radiation is utilized to build new biomass, which is indicated as "added to exergy storage" (exergy and eco-exergy will be explained in Chapter 3, but consider them measures of the amount of solar energy that is used for growth). The maintenance requires, of course, more energy for the mature stage, because it contains more biomass, which implies more energy is used in respiration.

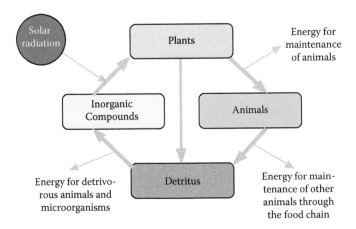

FIGURE 3.4 An ecosystem is a biochemical reactor. The input of energy comes from the solar radiation. The biologically important elements cycle and carry the energy, which is utilized by heterotroph organisms to support the life processes.

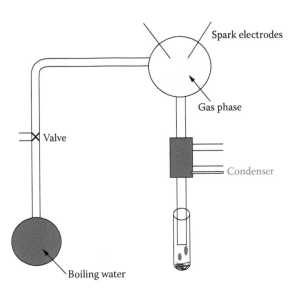

FIGURE 4.1 The apparatus used by Stanley Miller and Harold Urey to simulate reactions in the primeval atmosphere.

FIGURE 6.7 Danish beech forest, May 6. During a period of about 10 days the forest bursts into leaves. The biomass increases very rapidly. Growth form I is dominant.

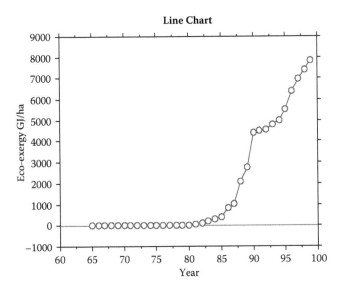

FIGURE 6.8 The development of eco-exergy on the island of Surtsey, a volcanic island formed south of Iceland by a volcanic eruption in 1962. The island is 150 ha and the eco-exergy for plants and birds has been found in GJ/ha from 1964 to 2000 (indicated as 100 on the graph). The eco-exergy is approximately increased exponentially; see also Figure 6.9.

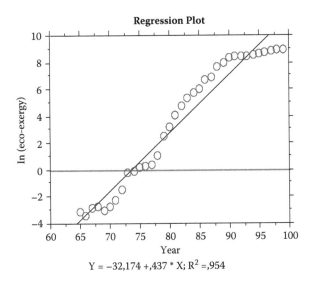

$$Y = -32,174 +,437 * X; R^2 =,954$$

FIGURE 6.9 Regression plot of log eco-exergy in GJ/ha (see Figure 6.8) vs. the years 1964–2000 (indicated on the graph as 100). As the regression plot with good approximation is linear with $R^2 = 0.954$, it is possible to conclude the eco-exergy on the island with good approximations has increased exponentially.

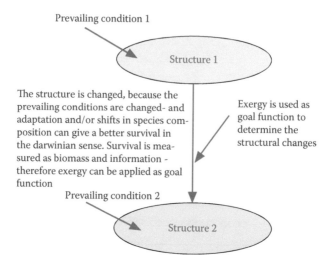

FIGURE 7.5 The theoretical considerations behind the application of SDM.

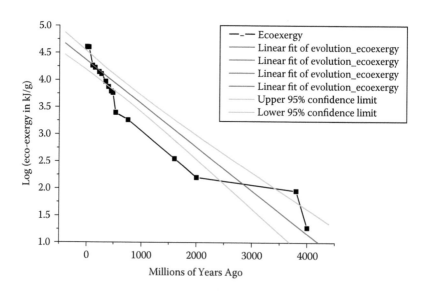

FIGURE 7.14 Semilogarithmic plot of eco-exergy density vs. time. The plot is with a rough approximation linear (the red line), which implies that the eco-exergy density has increased exponentially. The correlation coefficient is 0.95. The confidential intervals ± one standard deviation are indicated by the green lines, The increase, however, has been particularly fast since the Cambrian explosion and just after the emergence of the first primitive life.

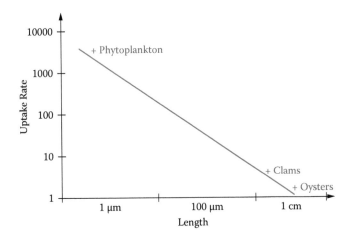

FIGURE 8.3 Uptake rate (µg/g 24 h) plotted against the length of various animals (CD): (1) phytoplankton, (2) clams, and (3) oysters (From Jørgensen, S.E., *Ecol. Model.*, 22, 1–12, 1984.)

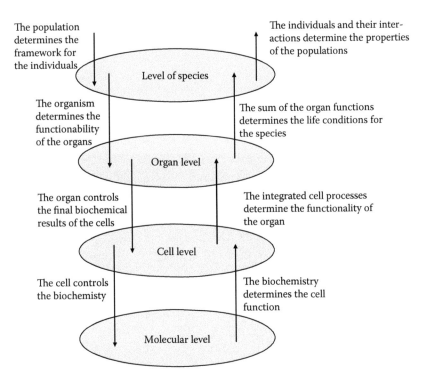

FIGURE 9.2 The four lower levels of the ecological hierarchy are presented. The interactions between the levels, including the fifth level, are indicated.

FIGURE 9.5 Beautiful landscape in Rocky Mountains, Canada. The landscape consists of a lake, a river (not shown in the figure), wetlands in the littoral zone of the lake, a forest, a mountain ecosystem below the timber line, and a mountain ecosystem above the timber line. All the ecosystems influence each other and are open.

FIGURE 10.2 One square meter of a natural ecosystem on an island in the Adriatic Sea. It was possible during a period of 2 h to count 92 plant and insect species.

FIGURE 10.3 The photo shows a shallow lake or wetland 13 km north of Copenhagen. The freshwater ecosystems in the temperate zone often have a very high diversity, and they offer many ecosystem services to society (see also Table 10.4).

FIGURE 11.4 Removal of planktivorous fish by a fishery in a North Italian lake.

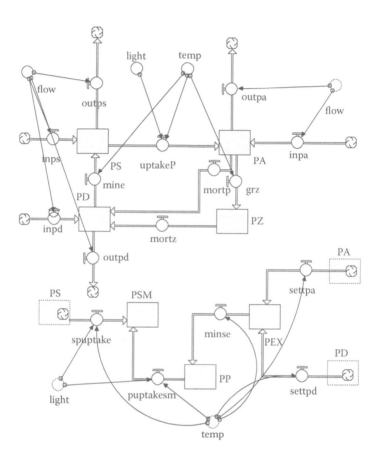

FIGURE 11.5 The conceptual diagram of the Lake Mogan eutrophication model focusing on the cycling of phosphorus. The model has seven state variables: soluble P, denoted PS; phosphorus in phytoplankton, PA; phosphorus in zooplankton, PZ; phosphorus in detritus, PD; phosphorus in submerged plants, PSM; exchangeable phosphorus in the sediment, PEX; and phosphorus in pore water, PP.

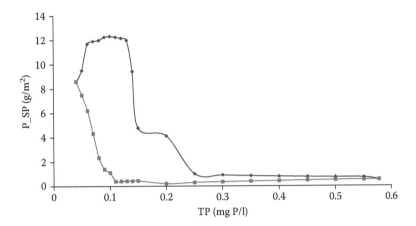

FIGURE 11.6 The graphs show the reaction of submerged plant phosphorus to an increase of phosphorus to about 400 μg/L (blue curve), followed by return to the original, about 80–85 μg/L (purple curve). When the phosphorus concentration is increased, P-SP increases due to the higher concentration of nutrients, but at about 250 μg/L the submerged vegetation disappears and is replaced by phytoplankton-P. At the return to 80–85 μg/L the submerged vegetation emerged at 100 μg/L. The hysteresis behavior is completely in accordance with Scheffer et al. (2001).

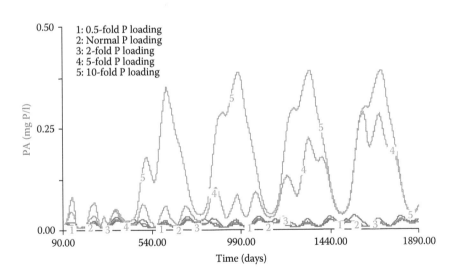

FIGURE 11.7 The phytoplankton-P concentration as a function of time is shown when the phosphorus concentration of the water flowing into the lake is increased by (1) a factor of 0.5 (blue), (2) normal (factor of 1.0, red), (3) a factor of 2 (purple), (4) a factor of 5 (green), and (5) a factor of 10 (orange). The present concentration of phosphorus in the lake is 80–85 μg/L. (1) to (3) give no changes, while (4) and (5) give a significant increase of phytoplankton, which becomes dominant.

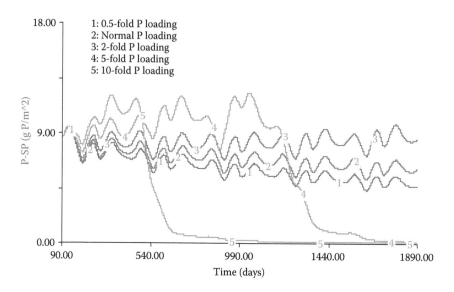

FIGURE 11.8 The submerged plant concentration as gP/m² as a function of time is shown when the phosphorus concentration of the water flowing into the lake is increased by (1) a factor of 0.5 (blue), (2) normal (factor of 1.0, red), (3) a factor of 2 (purple), (4) a factor of 5 (green), and (5) a factor of 10 (orange). The present concentration of phosphorus in the lake is 80–85 μg/L. (1) to (3) give no changes, while (4) and (5) first give a minor increase of the concentration, followed by a significant decrease, when phytoplankton becomes dominant; compare with Figure 11.6.

FIGURE 11.9 Clear water stage or structure by dominance of submerged vegetation.

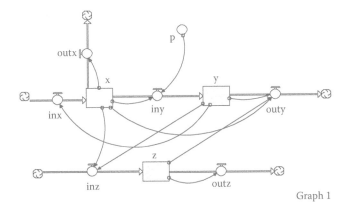

FIGURE 11.10 A simple model that shows chaotic behavior.

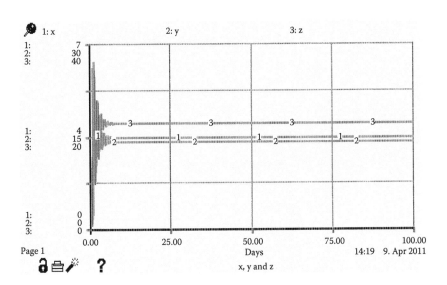

FIGURE 11.11 Simulations of the model in Figure 11.8, using iny = 25*x (see the equations in Table 11.1).

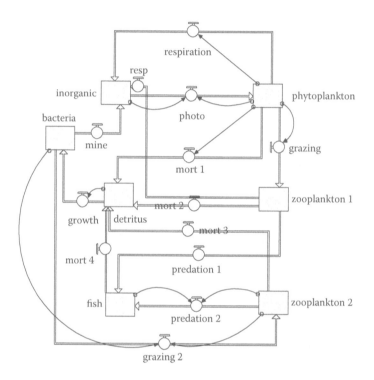

FIGURE 11.14 Model applied for the parameter examinations that resulted in Figures 11.15 and 11.16.

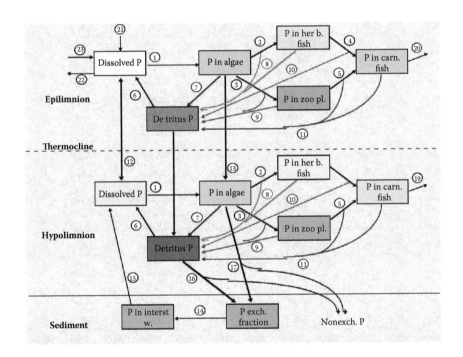

FIGURE 12.1 Deeper lakes have more comprehensive networks, because the lake is divided by the thermocline into epilimnion and hypolimnion.

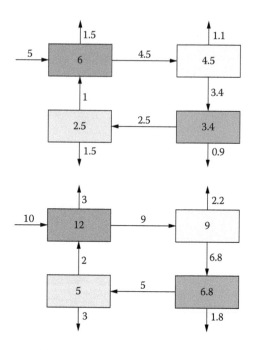

FIGURES 12.4 AND 12.5 The difference between the two figures is the input of eco-exergy or energy. The networks are at steady state and the flows are first order donor determined. The double input of eco-exergy or energy is able to support twice as much storage in all four compartments.

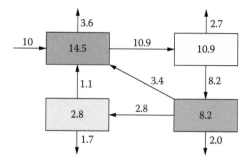

FIGURE 12.6 An extra cycling from compartment 3 to compartment 1. The result is that the eco-exergy, which is 32.8 in Figure 12.5, increases to 36.4.

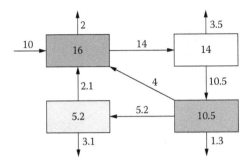

FIGURE 12.7 The first compartment represents bigger organisms than in Figure 12.6, which implies that the relative loss by respiration is reduced. Therefore, more eco-exergy is available to support the other compartment and recycle. The result is a total eco-exergy of 45.7.

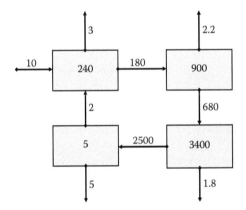

FIGURE 13.3 Same network as in Figure 12.5, but eco-exergy is applied instead of energy.

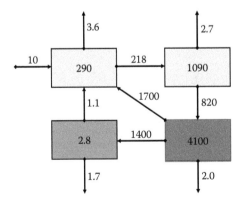

FIGURE 13.4 Same network as in Figure 12.6, but eco-exergy is applied instead of energy.

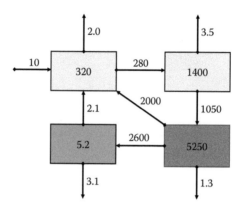

FIGURE 13.5 Same network as in Figure 12.7, but eco-exergy has replaced energy.

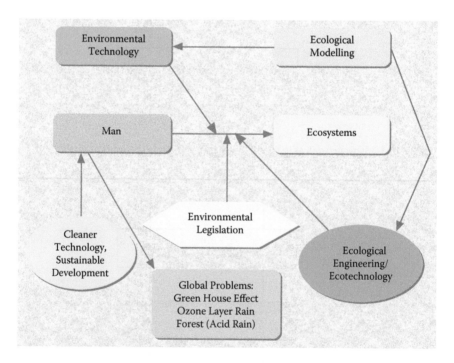

FIGURE 15.1 Conceptual diagram of the complex ecological-environmental management of today, where there are various toolboxes available to solve the problems and where the problems are local, regional, and global.

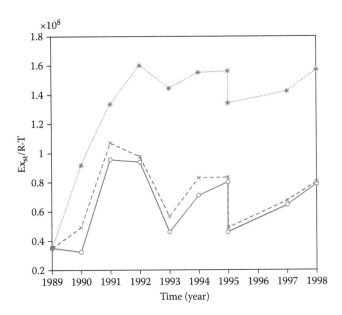

FIGURE 15.3 Eco-exergy mean annual values from 1989 to 1998: present scenario (continuous line), removal of *Ulva*, optimal strategy from cost-benefit point of view (dotted line), and nutrients load reduction from watershed (dashed line).

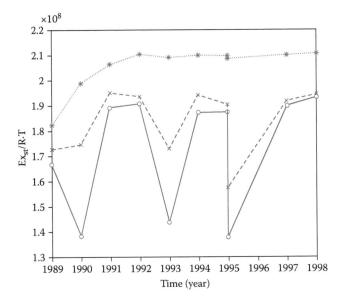

FIGURE 15.4 Specific eco-exergy mean annual values from 1989 to 1998: present scenario (continuous line), removal of *Ulva*, optimal strategy from cost-benefit point of view (dotted line), and nutrients load reduction from watershed (dashed line).

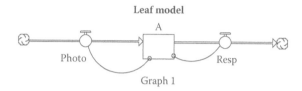

INTEXT FIGURE 3.1 Conceptual diagram of the leaf model.

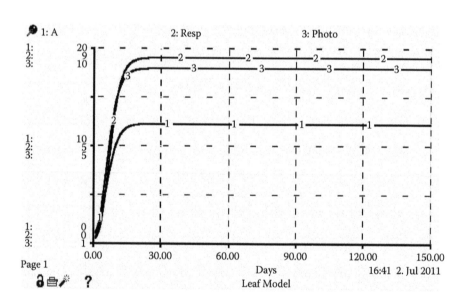

INTEXT FIGURE 3.2 Results of the leaf model.

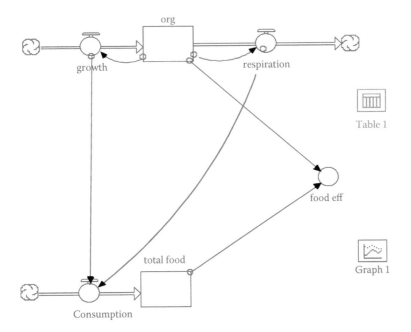

INTEXT FIGURE 7.1 Conceptual diagram of the horse evolution model.

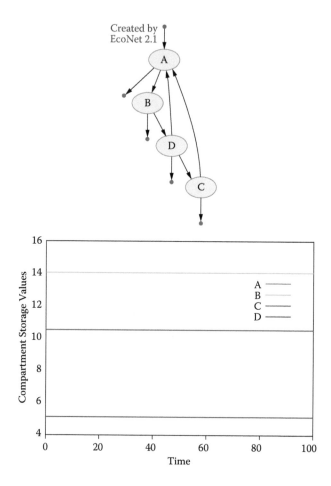

INTEXT FIGURE 12.1 Storage values as time.

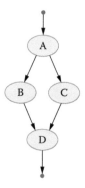

INTEXT FIGURE 12.2 The network.

Throughflow Analysis (N: z –> T)

–	Detritus	Microbiota	Meiofauna
Detritus	1	0	0
Microbiota	0.428571	1	0
Meiofauna	0.991597	0.980392	1

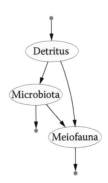

INTEXT FIGURE 12.3

Storage Analysis (S: z –> x)

–	Detritus	Microbiota	Meiofauna
Detritus	2.8574	0	0
Microbiota	0.840336	1.96078	0
Meiofauna	4.31129	4.26257	4.34783

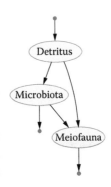

INTEXT FIGURE 12.4

Utility Analysis (U)

–	Detritus	Microbiota	Meiofauna
Detritus	**0.229593**	−0.470469	0.128122
Microbiota	−0.297411	**0.616033**	−0.220334
Meiofauna	0.270242	−0.43321	**0.157403**

Utility Relations (sign of U)

–	Detritus	Microbiota	Meiofauna
Detritus	+	−	+
Microbiota	−	+	−
Meiofauna	+	−	+

INTEXT FIGURE 12.5

5

The Biochemistry
of Ecosystems

Through analysis to the specific statement—
Through synthesis to the general statement.

The biochemistry of all organisms is approximately the same, which implies that the elementary composition of different types of organisms is also very much the same. The general composition and its variation and the first step toward the introduction of the biochemically based life processes are presented. The difference between prokaryote and eukaryote cells is discussed. The characteristics of biochemical processes lead to eight clear conditions for carbon-based life.

5.1 A General Biochemistry for Living Systems

The biochemistry for primitive cells and that for the most advanced animals, mammals, are surprisingly similar. The metabolism follows approximately the same pathways. The core processes for photosynthesis are also the same for all plants. It is therefore not surprising that the elementary composition of various organisms is very similar.

The wet weight composition may be different because the water content varies more than the dry matter composition from organism to organism. A typical dry matter content of plants may be 12.5%. This means that the concentrations of various elements on a wet basis given in Table 2.2 can be changed to the dry matter composition (see Table 5.1). The compositions given in this table are with good approximations applicable to all plants and even to animals.

Table 5.1 has listed the 19 elements that are generally found in all organisms. A handful of elements can be found in addition to the 19 listed at low concentrations; for instance, iodine and fluorine, and a few trace elements have been found to be characteristic in just a few species: selenium, nickel, vanadium, chromium, and even the toxic element cadmium may substitute into enzymes that typically use zinc. In principle, each of the applied elements can be a limiting factor for growth, as discussed in

TABLE 5.1   Average Elemental Composition
of Freshwater Plants, Dry-Weight Basis

Element	Plant Content (%)
Oxygen	11.1
Hydrogen	1.4
Carbon	45.5
Silicon	9.1
Nitrogen	4.9
Calcium	2.8
Potassium	2.1
Phosphorus	0.56
Magnesium	0.49
Sulfur	0.42
Chlorine	0.42
Sodium	0.28
Iron	0.14
Boron	0.007
Manganese	0.0049
Zinc	0.0021
Copper	0.0007
Molybdenum	0.00035
Cobalt	0.000014

Section 2.2. A limiting element must of course be absolutely necessary for the considered organism and be present in a concentration that is lower than the other important elements relative to its use for building new biomass for the focal organism. The composition of the ecosphere reflects, however, the composition of the organisms, which means that the elements are often present in a ratio close to the concentrations in Table 5.1. Furthermore, the composition of organisms is not an exact unchangeable value but rather a range. For instance, most plants can manage to grow with a phosphorus content of 0.4–2.0%. If the environment has a high phosphorus concentration relative to the need of the organisms, the organisms will accumulate more phosphorus, and if the phosphorus concentration of the environment is low, the plants are able to adapt to the conditions and cope with less phosphorus. This adaptability of the plants to the composition of the environment is reflected in the range of concentrations in the organisms and implies that a limiting concentration of an element is not a sharp exact value, but is a range that gradually, at lower concentrations, will limit the growth more and more. See also the discussion in Section 2.3.

Approximate ranges for C, N, and P are given for different organisms in Table 5.2 to show (1) the variability among species and (2) the possible ranges. The differences among species are not very pronounced, while the ranges may be relatively wide for phosphorus, which they are generally for elements with low biochemical concentrations.

TABLE 5.2 Approximate Ranges of C, N, and P
Concentrations as % Dry Matter for Different Organisms

Organisms	%C	%N	%P
Terrestrial plants	36–64	0.3–6.4	0.02–1.0
Benthic invertebrates	35–57	6–12	0.2–1.8
Terrestrial insects	36–61	7–12.5	0.5–2.5
Birds and mammals	32–60	6–12	0.7–3.7
Fish	38–52	7–12	1.5–4.5
Zooplankton	35–60	7–12.5	0.5–2.5
Phytoplankton	35–60	5–12	0.5–2.5

5.2 The First Steps of the Evolution toward a Biochemistry

It cannot be excluded that the first primitive cells came from space, but whether first biochemistry was developed on the earth or in space, we should somehow try to explain it. Developments over the past decade have given new credibility to the idea that earth's biosphere could have arisen from an extraterrestrial seed. Planetary scientists have learned that our solar system previously included many worlds with liquid water, which we consider is the essential ingredient for carbon-based life as we know it. Life may have evolved on Europa (Sweinsdottir, 1997), Jupiter's fourth largest moon, or on Saturn's biggest moon, Titan, which is rich in organic components. Biologists have furthermore discovered microorganisms durable enough to survive at least a short journey inside a meteorite.

Life processes are based on proteins with a certain and specific amino acid sequence, which it is necessary to remember to be able to start and continue the evolution. We therefore need the genes and the proteins simultaneously, because the right amino acid sequence cannot be transferred to the next generation without the genes, and the genes would have no function without the proteins that control the life processes. We cannot today imitate the first biochemical processes—not yet at least—but a possibility could be that RNA has been formed. The process should not be too difficult as fragments of these molecules are formed from inorganic molecules by an input of energy. RNA has an ability to replicate and adsorb specific amino acids to the four amino bases that are the reactive parts of RNA. Probably these first contacts between RNA and amino acids took place on clay particles or at the surface of scum. RNA should therefore be able to build various polypeptides, and they would contain the 20 amino acids that we know make up the life-determining proteins. In this context, it is important that a successful amino acid sequence, i.e., polypeptides or proteins with an amino acid sequence, which could serve as enzymes for useful conversions of other organic molecules, could be remembered. An obvious question would be: Why are proteins made from 20 amino acids species that serve universally for their synthesis? All the amino acids that are used for protein syntheses are in the L-form—the D-forms are never applied. It is not a question of the relative abundance. A likely answer to this riddle is that the amino acids were selected for protein syntheses due to their ability to interact with RNA. The probability that the L-form was selected was simply 50%, but once selected it was not changed.

The first cells were much simpler than present-day cells. They are often denoted protocells. They were, however, subject to mutations. These affected the efficiency of the RNA, the resilience of the coding system, and the quality of the synthesized proteins with respect to their ability to favor protocell growth and proliferation. De Duve (2002) presumes that the first proteins in an organism, which may not even have a cell membrane, were only 20 amino acids long, which implies a biological eco-exergy of $RT \ln 1/20^{-20} = 8.34 \times 300 \times 20 \times 3 = 150\,120$ J/mole (see Chapter 4), or as the molecular weight of 20 amino acids is about 3,000 g, the eco-exergy in kJ/g becomes 0.05 kJ/g. This gives a β-value; see Table 3.1 on $(18.7 + 0.05)/18.7 = 1.0027$, or less than the β-value for virus. The eco-exergy density can be calculated to be $18.7 \times 1.0027 = 18.75$ kJ/g.

The proteins with a chain of 20 amino acids have 20^{20}, or about 10^{26} possible arrangements. De Duve calculated (2002) that these amounts of arrangements with even 99.9% of the volume to spare would only require a volume of $20 \times 50 \times 0.1 = 100$ km^3. The experiment to find the best amino acid sequence takes place on the molecular level—nature uses super-nanotechnology. De Duve presumes that out of the 10^{26} different arrangements only 1,000 were favorable in the sense that they were able to utilize organic matter for the metabolic processes and to favor growth and proliferation. These probability calculations seem to explain the start of the biochemistry.

The next step is of course that the proteins react with each other and form proteins with bigger molecular weights. Viruses can code for 2,000 amino acids, and an organism similar to a virus may have been the next step. A virus has a β-value of 1.01. Shortly (10^8 years?) after the earth attained a suitable temperature on its surface, probably about 4.0 billion years ago, the life forms were probably as described above: a combination of RNAs and proteins that could reproduce and metabolize. The exergy density is $18.7 \times 1.01 = 18.9$ kJ/g.

The composition of the organisms reflects, of course, the organic compounds that are general building blocks for living organisms. The stoichiometries of important organic compounds that are used as building blocks are given in Table 5.3.

Compare the three tables, Tables 5.1–5.3, to see how the elementary composition of a living organism is determined by the building blocks.

TABLE 5.3 Approximate Percentage Composition of Selected Macromolecules and Other Organic Compounds That Are Biochemical Building Blocks

Compound	%C	%N	%P	%O	%H	%S
Protein	54	17	0	20	7	2.7
Collagen	54	1,810	0	21	7	0.5
DNA	38	17	10	31	4	0
RNA	36	16	10	35	3	0
ATP	24	14	18	41	2	0
Cholesterol	84	0	0	4	12	0
Cellulose and starch	46	0	0	51	3	0
Lignin	63	0	0	31	6	0
Chitin	44	7	0	42	7	0
Chlorophyll	74	6	0	9	8	0

5.3 The Prokaryote Cells

The three important steps of very primitive life toward the prokaryote cell are:

1. DNA was formed from RNA. DNA covers the information storage and replication, while RNA utilizes the information in protein synthesis. This division made it possible to regroup the genes, and the replication of the genes could be carried out in a synchronous fashion, coordinated with cell division.
2. A cell membrane was formed to protect the life processes, the RNA, DNA, and proteins (enzymes). Physically, biological membranes are not very different from soap bubbles. They are very thin, highly flexible, self-sealing films that consist of mainly carbon and hydrogen. Their constituent molecules are phospholipids. Formation of the cell membrane was major progress. It became possible to protect the biochemical components making up the life processes, to maintain more easily another composition inside the cell than outside, but still mediate exchanges with the outside.
3. Introduction of ATP to facilitate the exchange of energy between different parts of the cell and between different biochemical processes. ATP is relatively easy to produce from other organic molecules by input of energy. It can therefore not be excluded that ATP participated in the biochemical processes of the very first primitive life consisting of small proteins and RNA, as described above.

The development corresponding to these three points has probably taken on the order of 100–200 million years—a very long time, particularly compared with the later steps of evolution. The oldest fossils of cells are about 3.8 billion years old and were found in Greenland (Haugaard Nielsen, 1999). The minimal cell has a β-value of about 5.0 (see Table 3.1). The exergy density is now as high as $18.7 \times 4.88 = 91$ kJ/g. The free energy flow density of synthesizing organic polymer in primitive cells (Geigy, 1990) is on the order of 0.02 J/s kg.

The primeval world was oxygen-free and remained so until about 2.3 billion years ago. The level of atmospheric oxygen started rising at that time due to introduction of photosynthesis and, a few hundred million years later, reached values compatible with aerobic life. The organism responsible for this significant change in the composition of the atmosphere was *cyanobacteria*. Oxygen is deadly toxic to anaerobic organisms, but new life forms—aerobic microorganisms—were able to cope with the new challenge of oxygen.

The time from about 3.5 to 2.0 billion years ago gave rise to a wide spectrum of different prokaryote cells using different biochemical processes, particularly to oxidize the organic matter that is the energy source of these cells. Furthermore, the biochemical processes included the metabolic processes and became more refined. Some scientists talk about bacteria as superstars of the living world, because of the high reproduction rate. A β-value of 8.5 (see Table 4.1) represents the prokaryote cell in the most developed stage, which was reached approximately 2 billion years ago. The corresponding exergy density is $18.7 \times 8.5 = 159$ kJ/g.

5.4 The Eukaryote Cells

The eukaryote cells are much larger than prokaryotes. They are organisms of considerable complexity. They are composed of many different parts, endowed with distinct functions. They have a highly organized pulp, denoted cytoplasm, and a central kernel, called the nucleus. Mitochondria contain the respiratory enzymes of the cells. Whereas the cytoplasm is the site of the majority of biochemical processes, including the protein synthesis, the nucleus is taken care of by the genetic operations. Eukaryotes have a well-organized system for duplicating their DNA exactly into two copies during cell division. This process, denoted mitosis, is much more complex and precise than the simple splitting found in prokaryotes. Eukaryotes can perform sexual reproduction in which the DNA of two cells is shuffled and redealt into new combinations. It has increased the number of combinations that can be tested per unit of time for better fitness enormously. The almost infinite variety possible from sexual reproduction provides for change in a changing world. Because of the constraints from changes of the life conditions, a successful genotype that would pass unaltered into the next generation would not be able to continue its success. Relying solely on asexual reproduction, the capacity of species to adapt genetically is very limited; once its phenotypic flexibility has been exceeded, there is little prospect of accommodating further changes. Sexual reproduction may be risky, but it does at least produce new combinations of genes and a possibility of increasing the range of tolerance. The costs of sexual reproduction are high, and so is the risk of failure. Each individual has to meet a partner, and two gametes, the sperm and the egg, must fuse to form a single cell, the zygote. The zygote must have a viable combination of genes, always with the risk that the new genotype may be unable to support the development of a new individual. To survive, each offspring needs to be well matched to the environment in which it will grow and reproduce. When a large number of offspring is produced, all are slightly different from one another, and there is therefore an increased chance that some will survive in the changing world. There is, however, wastage, and there is a major cost to the parents. Some combinations, some new genotypes will inevitably be less fit than their parents. Asexual reproduction avoids all these risks and costs and may be more beneficial in a not changing environment. So, the constraints due to the changing life conditions can actually be considered the prerequisite for evolution.

The gradual accumulation of changes resulting from mutations and sexual recombinations and a few other mechanisms over many generations produces new species. The cells are surrounded with a great diversity of outer coverings of enormous complexity and made of a wide variety of substances, mostly proteins, carbohydrates, and polymers of combinations of both. Finally, it should be mentioned that eukaryotes typically are 10 times larger in diameter or 1,000 times larger in volume than prokaryotes. Had the transformation from prokaryotes to eukaryotes not taken place, the living world of today would still have been only bacteria.

Any microbiology textbook gives a detailed description of the relatively significant differences between the prokaryote and the eukaryote cells. The differences are summarized in Table 5.4. The development from prokaryotes to eukaryotes started probably about 2.3 billion years ago at the time when the oxygen concentration of the atmosphere started to rise (de Duve, 2002). Fossils of eukaryotes have been dated to 2,100 million

TABLE 5.4 The Most Significant Differences between Prokaryotes and Eukaryotes

	Prokaryotes	Eukaryotes
Size	1–10 μm	10–100 μm
Nucleus	None; the chromosomal region is called the nucleolus	Nucleus separated from cytoplasm by nuclear envelope
Intracellular organization	Normally, no membrane-separated compartments and no supportive intracellular framework	Distinct compartments, e.g., nucleus, cytosol with cytoskeleton, mitochondria, endoplasmic reticulum, Golgi complex, lysosomes, plastids
Gene structure	No introns; some polycistronic genes	Introns and exons
Cell division	Simple	Mitosis or meiosis
Ribosome	Large 50S subunit and small 30S subunit	Large 60S subunit and small 40S subunit
Reproduction	Parasexual recombination	Sexual recombination
Organization	Mostly single cellular	Mostly multicellular, with cell differentiation

Source: After Klipp, E., et al., *Systems Biology in Practice. Concepts, Implementation and Application*, Wiley-VCH Verlag GmbH, Weinheim, Germany, 2005, p. 21.

years ago (Madsen, 2006). The mitocondria that are the chemical factory of the eukaryote cells were probably formed as a result of one cell penetrating another cell, starting a biochemical cooperation.

There is strong evidence that the ancestors of eukaryotes were anaerobic, although eukaryotes still may be divided into anaerobic and aerobic organisms, which indicates that the first phase in the development may have taken place before oxygen in the atmosphere started to rise (see the discussion in de Duve, 2002).

Typical early eukaryote cells are yeast and diatoms that represented the most developed life forms about 1.5–1.8 billion years ago. The step forward in evolution from prokaryote to eukaryote is enormous, and most probably archaea could be considered the link between prokaryote and eukaryote with β-values of about 13–14. Eukaryotes have β-values of at least 18–20. This implies exergy densities of about $18.7 \times 19 = 355$ kJ/g on average.

The size of cells and cell division are determined by exchange of energy and matter between the cell and the environment. The uptake of substrate from the environment by a spherical cell is proportional to the surface area of the cell $4\pi r^2$, where r is the radius of the sphere. The transport from the surface to the cell takes place by a fast active transport, and the concentration at the surface is therefore 0. The flux of matter toward the cell surface can be considered constant, which implies that the concentration gradient will decrease with the distance from the cell center, dis, in the exponent –2: $dC/ddis = k\ dis^{-2}$. This means that the concentration at the distance dis can be found after differentiation to be $C(dis) = C(1 - r/dis)$. The uptake rate at the surface is the gradient at the surface = C/r times the area, and therefore the uptake rate becomes:

$$\text{Uptake (mol/s)} = 4\pi\ C\ D\ r \tag{5.1}$$

where D is the diffusion coefficient and C is the concentration of substrate in the environment. The amount of substrate taken up should correspond to the need, which is proportional to the volume of the cell and can be expressed by the following equation:

$$\text{Need (mol/s)} = f\,4\,\pi\,r^3/3 \tag{5.2}$$

where f corresponds to the need per unit of volume. By these equations it is possible to find the max value of r, denoted r_{max}. If r is $> r_{max}$, the need is bigger than the uptake rate. The maximum cell size will correspond to a balance between the need and the uptake. It can therefore be found as $(3D\,C/f)^{0.5}$.

The entropy production in a cell is proportional to the metabolism and is therefore proportional to the volume, while the outflow of entropy is proportional to the surface. If the cell grows too big, the outflow is too small to compensate for the entropy production and the cell will die. As a result, the cell is divided into two cells and thereby the volume is maintained and the surface area increases. Cell division makes it possible to grow and still maintain the entropy balance.

Example 5.1

The oxygen concentration in freshwater at 16°C is by equilibrium with the atmosphere about 0.3 mol/m³. Active bacteria have a metabolism requiring about 21 moles of oxygen/m³s. Oxygen is limiting for the activity and has a molar diffusion coefficient of $2.1*10^{-9}$ m²/s. What would be the size of active freshwater bacteria?

Solution

Maximum cell radius = $(3\,D\,C/f)^{0.5} = (3*2.1*10^{-9}*0.3/21)^{0.5}$
$= 9.5$ mm

5.5 The Temperature Range Needed for Life Processes

The input of energy for ecosystems is in the form of the solar photon flux. This comes as small portions (quanta) of energy (= hv, where h is Planck's constant and v is the frequency), which implies that the exergy at first can only be utilized at molecular (lowest) levels in the hierarchy. The appropriate atoms or molecules must be transported to the place where order is created. Diffusion processes through a solid are extremely slow, even at room temperature. The diffusion of molecules through liquids is about three orders of magnitude faster than in solids at the same temperature. Diffusion coefficients for gases are ordinarily four orders of magnitude greater than for liquids. This implies that the creation of order (and also the inverse process, disordering) is much more rapid in liquid and gaseous phases than in solids. The temperature required for a sufficiently rapid creation of order is consequently considerably above the temperature of open space, 2.726 K. As far as diffusion processes in solids, liquids, and gases are concerned, gaseous diffusion allows the most rapid mass transport. However, many molecules on earth that are necessary for ordinary carbon-based life do not occur in a gaseous phase, and liquid diffusion, even though it occurs at a much slower rate, is of particular importance for biological ordering processes.

The diffusion coefficient increases significantly with temperature. For gases, the diffusion coefficient varies with temperature approximately as $T^{3/2}$ (Hirschfelder et al., 1954), where T is the absolute temperature. Thus, we should look for systems with the high-order characteristic of life, at temperatures considerably higher than 2.726 K. The reaction rates for biochemical anabolic processes (biochemical formation of complex biologically important compounds as, for instance, enzymes and hormones) are highly temperature dependent (see Straskraba et al., 1997). The influence of temperature may be reduced by the presence of reaction-specific enzymes, which are proteins formed by anabolic processes. The relationship between the absolute temperature, T, and the reaction rate coefficient, k, for a number of biochemical processes can be expressed by the following general equation (see any textbook in physical chemistry):

$$\ln k = b - A/R^*T \tag{5.3}$$

where A is the so-called activation energy, b is a constant, and R is the gas constant. Enzymes are able to reduce the activation energy (the energy that the molecules require to perform the biochemical reaction). Similar dependence of the temperature is known for a wide spectrum of biological processes, for instance, growth and respiration. Biochemical and biological kinetics therefore point toward ecosystem temperatures considerably higher than 2.726 K.

The high efficiency in the use of low-entropy energy at the present room temperature on earth works hand in hand with the chemical stability of the chemical species characteristic of life on earth. Macromolecules are subject to thermal denaturation. Among the macromolecules proteins are most sensitive to thermal effects, and the constant breakdown of proteins leads to a substantial turnover of amino acids in organisms. According to biochemistry, an adult man synthesizes and degrades approximately 1 g of protein nitrogen per kg of body weight per day. This corresponds to a protein turnover of about 7.7% per day for a man with a normal body temperature. A too high temperature of the ecosystem (more than about 340 K) will therefore enhance the breakdown processes too much.

A temperature range between 260 and 340 K seems from these considerations the most appropriate to create the carbon-based life that we know on earth.

An enzymatic reduction of the activation energy makes it possible to realize basic biochemical reactions in this temperature range, without a too high decomposition rate, which would be the case at a higher temperature. In this temperature range anabolic and catabolic processes can, in other words, be in a proper balance.

5.6 Natural Conditions for Life

The conditions for creation of life, ordering processes out of disorder (or more specifically, chemical order by formation of complex organic molecules and organisms from inorganic matter), can now be deduced from the first, second, and third laws of thermodynamics, which are presented in Chapters 2–4. The four conditions are a summary of what is already presented in these three chapters, but below are added the biochemical conditions to complete the list of conditions for life and therefore ecosystems:

1. It is necessary that the system be open (or at least nonisolated) to exchange energy (as well as mass) with its environment.
2. An influx of low-entropy energy or exergy, that can do work, is necessary.
3. An outflow of high-entropy energy (heat produced by transformation of work to heat) is necessary (this means that the temperature of the system inevitably must be greater than 2.726 K).
4. Entropy production (or consumption of exergy) accompanying the transformation of energy (work) to heat in the system is a necessary cost of maintaining the order.

We can now add a fifth condition based on the influence of the temperature on biochemical processes:

5. Mass transport processes at a not too low rate are necessary. This implies that the liquid or gaseous phase must be anticipated. A higher temperature will imply a better mass transfer, but also a higher reaction rate. An increased temperature also means a faster breakdown of macromolecules, and therefore a shift toward catabolism. A temperature approximately in the range of 260-340 K must therefore be anticipated for carbon-based life.

The rates of biochemical reactions on the molecular level are determined by the temperature of the system and the exergy supply to the system. Hierarchical organization ensures that the reactions and the exergy available on the molecular level can be utilized on the next level, the cell level, and so on throughout the entire hierarchy: molecules → cells → organs → organisms → populations → ecosystems. The maintenance of each level is dependent on its openness to exchange of energy and matter. The rates in the higher levels are dependent on the sum of many processes on the molecular level. They are furthermore dependent on the slowest processes in the chain: supply of energy and matter to the unit → the metabolic processes → excretion of waste heat and waste material. The first and last of these three steps limit the rates and are determined by the extent of openness, measured by the area available for exchange between the unit and its environment relative to the volume. These considerations are based on allometric principles, which will be presented in Chapter 8.

In addition to the five conditions given above, it is necessary to add a few biochemically determined conditions. The carbon-based life on earth requires first of all an abundant presence of water to deliver the two important elements hydrogen and oxygen, as solvent for compounds containing the other needed elements (see below), and as a compound that is liquid at a suitable temperature with a suitable diffusion coefficient; a suitable specific heat capacity to buffer temperature fluctuations; and a suitable vapor pressure to ensure a suitable cycling (purification) rate of this crucial chemical compound.

Life on earth, as presented in Section 5.1, is characterized by about 25 elements. Some of these elements are used by life processes in micro amounts, and it cannot be excluded that other elements could have replaced these elements on other planets somewhere else in the universe. Several metal ions are, for instance, used as coenzymes and are often important parts of high molecular organic complexes. Other ions may be able to play

similar roles for biochemical processes and complexes. It is, on the other hand, difficult to imagine carbon-based life without at least most of the elements used in macro
amounts, such as nitrogen for amino acids (proteins—the enzymes) and amino bases,
phosphorus for ATP and phosphorous esters in general, and sulfur for formation of
some of the essential amino acids.

The additional biochemically determined conditions (constraints) can therefore be
summarized in the following two points:

6. Abundant presence of the unique solvent water is a prerequisite for the formation
   of life forms similar to the life forms we know from the earth.
7. The presence of nitrogen, phosphorus, and sulfur and some metal ions seems
   absolutely necessary for the formation of carbon-based life. We will touch further
   on this issue in the next chapter.

A last and eighth condition should be added: the seven other conditions should be maintained within reasonable ranges for a very long period of time. The genes may ensure that if
an advantageous property of an organism has been developed, the property can be inherited and the following generations will be able to maintain the advantageous property.
The probability to create (complex) life spontaneously is so low that even the time from
the big bang would not have been sufficient. It is therefore necessary that the development
toward life is made step-wise with conservation of each achieved progress to allow further
development to ride on the shoulders of the already made progress. Many mechanisms are
probably involved in the emergence of a progressive property on the first hand, but indis-
putable random processes based on trial and error are also important in the emergence of
progressive properties. This implies that carbon life is not formed overnight. The history of
evolution on earth shows that a suitable temperature and abundant water, probably on the
order of 10^8 years or more (Madsen, 2006), were needed to form, from inorganic components dissolved in water, the first living cells with some type of primitive genes to ensure a
continuous development (evolution). Fossils after phytoplankton have recently been found
at Isua, Greenland, by Minik Rosing (Madsen, 2006). The age of the fossils was determined
to be 3.8 billion years old, or about 100 million years after the termination of the massive
bombardment of meteors that characterized the first 600–700 million years after the earth
was born. Numerous theories have been published to explain how this development may
have happened, probably in many steps: inorganic matter formed organic molecules by a
throughflow of low-entropy energy (compare with Figure 4.1), organic molecules formed
high molecular organic compounds, self-catalytic processes occurred, complex organic
molecules were brought randomly in contact by adsorption on clay particles, and many
other processes, mentioned in the presented theories. Which of the theories is right is not
important in this context. The focal point is that the seven above-mentioned conditions
must be fulfilled for a sufficiently long period of time, which leads to the eighth condition:

8. As the formation of life from inorganic matter requires a very long time, probably
   on the order of 10^8 years or more, the seven conditions have to be maintained in
   the right ranges for a very long time, which probably exceeds about 10^8 years.

After the Mars Pathfinder mission, it was discussed whether Mars hosts or has hosted
life. Clearly, conditions 1–7 are not met on Mars today. The climate is too harsh and

water is far from being present in the amount needed for the planet to bear life. There are, however, many signs of a warmer and wetter climate at an earlier stage. It looks, therefore, like the seven conditions may have been valid, and the question is: Did they also prevail for a sufficient period of time? If later missions to Mars show that is the case, the next obvious question is: Will life inevitably be the result of self-organizing processes, if the eight conditions (the eighth condition about sufficient time should, of course, be included) are fulfilled? It should be expected that primitive life was present on Mars at an early stage, provided that the warmer and wetter conditions prevailed for sufficient time. The further evolution from (maybe prokaryotic) unicellular organisms to more and more complex organisms, as we know from earth, could not be realized on Mars, because the climate changed and the water disappeared. Latest investigations of Mars-originated meteorites have made it almost certain that there has previously been microbiological life on Mars. The latest geological investigations have furthermore shown that there has previously been plenty of water on Mars, which also points toward a prior existence of life on Mars. It is, of course, still an open question if this microbiological life is still present. The Mars Pathfinder mission will probably be able to answer this question.

Another possibility for life in our solar system exists on Europa, one of the moons of Jupiter (Sweinsdottir, 1997). Europa is characterized by a covering of ice. This implies that there is plenty of water on Europa, which means that one of the important conditions for life is fulfilled. Some researchers (Sweinsdottir, 1997) believe the chance of finding life on Europa is higher than on Mars. Europa has, of course, much less sunlight, and the surface temperature is probably too low, but volcanic activity in the deeper parts of the oceans on Europa is very probable, and it could provide the needed low-entropy energy for the formation and maintenance of life.

> The eight conditions presented above are all needed to support life and are prerequisites for ecosystems.

The question is whether the eight conditions are also sufficient to create ecosystems. We know that life is formed even under extreme conditions on the earth, but it will probably not be possible to answer the question before we have found life outside the earth. It can be shown (see Morowitz, 1968) that a flow of energy through a system will create order, for instance, by creating at least one cycle of energy and matter. The next chapter will go into detail about how an inflow of low-entropy energy or exergy that can do work will move a system away from thermodynamic equilibrium—create more structure and more order. It can be considered a natural law, but it is of course another question, whether the structure and order actually is life. It will require all eight conditions.

5.7 Ecological Stoichiometry

Biochemistry is determined by the evolution of the first 2 billion years or so, where the development was stepwise from simple organic molecules to more complex organic molecules to very complex organic polymolecules that formed the first life forms, to the prokaryote and to the eukaryote cell. Numerous biochemical reactions describe the decomposition of organic matter utilized as a source of energy: formation of proteins controlling the life processes as enzymes; formation of important biochemical molecules as, for instance, hormones and energy carriers like ATP and coenzymes; formation of carbohydrates from carbon dioxide and water by photosynthesis; and formation of structural compounds forming cell membranes, skin cells, teeth, bones, and so on.

All the processes can be described as chemical processes following the stoichiometry, or following it at least approximately. This implies that the processes of the organisms and their cells follow the stoichiometry formulated in chemistry. Equation (3.7) was an example:

$$C_{3,500}H_{6,000}O_{3,000}N_{600} + 4,350\ O_2 \rightarrow 3,500\ CO_2 + 2,700\ H_2O + 600\ NO_3\text{-} + 600\ H^+$$

This equation describes the decomposition of a typical high molecular organic molecule representing, for instance, detritus. A molecular weight of 102,400 is presumed as a typical representative for the high molecular organic molecules forming organic matter.

The general biochemistry means furthermore that the elementary composition as presented in 5.1 becomes general for all organisms and for the biochemistry of all living components of ecosystems, as expressed in the Redfield ratio, C:N:P = 48:7:1 on a weight basis, while the mole basis becomes with approximate ratio 120:16:1.

The rate of cycling limiting elements is critical for ecosystems. The cycling implies transfer of elements between biotic and abiotic components of the ecosystem. Autotrophs take up resources from their environment, and they also leak nutrients back to the environment, for instance, to soil or water. They also produce litter, which is decomposed and thereby nutrients are released. Heterotroph consumers participate in the cycling by ingesting food and releasing waste back to the environment. Generally the process food/resources → organism + waste follows the mass conservation principle. Therefore, if the elementary composition of food/resources and of the organisms is known, we can find the elementary composition of the waste. It is assumed that individual species have strict homeostasis using optimum stoichiometric coefficients in their chemical makeup. This implies that the scarcest element relative to its need will be better retained by the organism. For example, food of a relatively low P content is ingested by a homeostatic consumer. Required P is retained with elevated efficiency, and the waste will contain relatively less P than the food/resources. As a consequence, it has been found that when N:P is increasing in the food, homeostatic consumers are disposing of the excess nutrient and the nutrient that is most limiting is retained for its own needs.

Example 5.2

A food source has the C:N:P ratio 100:5:0.5, while the organism living on the food source has the C:N:P ratio 45:7:0.8. When the food is utilized by the organism by an efficiency of 60%, what is the ratio of the waste?

Solution

105.5 parts of C + N + P of the food source can be utilized by the organism as 60 parts C, 3 parts N, and 0.3 parts P. It means that P is limiting the growth of the organism and is used up before N and C. To obtain 0.8 parts of P, 8/3 as much matter has to be utilized out of the 105.5 parts, or 8 parts of nitrogen and 160 parts of C. Only 7 parts are needed of nitrogen, which means that 1 N is in surplus, and only 45 parts of C are needed out of 160, which means that 125 parts are in surplus. The conclusion is that the waste will contain C:N:P in the ratio 125:1:0.

5.8 Summary of Important Points in Chapter 5

1. The biochemistry of different living organisms is surprisingly similar. This implies that the chemical composition of different organisms is also similar for the most common 20 elements in biological material. Proteins, carbohydrates, and fat are the three main classes of organic compounds.
2. The complexity of prokaryote and eukaryote cells is very different, although they still have almost the same biochemistry and elementary composition.
3. The general biochemistry and elementary composition of living matter make it possible to make quantitative calculations based on ecological stoichiometry.
4. The necessary eight conditions for carbon-based life are (they may also be sufficient, but this is not yet possible to conclude):
   a. It is necessary that the system is open (or at least nonisolated) to exchange energy (as well as mass) with its environment.
   b. An influx of low-entropy energy or exergy, that can do work, is necessary.
   c. An outflow of high-entropy energy (heat produced by transformation of work to heat) is necessary (this means that the temperature of the system inevitably must be greater than 2.726 K).
   d. Entropy production (or consumption of exergy) accompanying the transformation of energy (work) to heat in the system is a necessary cost of maintaining the order.
   e. Mass transport processes at a not too low rate are necessary (a prerequisite). This implies that the liquid or gaseous phase must be anticipated. A higher temperature will imply a better mass transfer, but also a higher reaction rate. A temperature approximately in the range of 260–340 K must therefore be anticipated for carbon-based life.
   f. Abundant presence of the unique solvent water is a prerequisite for the formation of life forms similar to the life forms we know from the earth.
   g. The presence of nitrogen, phosphorus, sulfur, and some metal ions seems absolutely necessary for the formation of carbon-based life.
   h. As the formation of life from inorganic matter requires a very long time, probably on the order of 10^8 years or more, the seven conditions have to be maintained in the right ranges for a very long time, which probably exceeds 10^8 years.

Exercises/Problems

1. The higher the N:P ratio in an aquatic ecosystem, the faster N is recycled relative to P. The following results have been obtained:

N/P Molar Ratio in the Ecosystem	N/P Ratio of Recycling Rate (moles/24 h)
10	9
20	23
30	41
40	80

Explain why the recycling rate is increasing by an increasing ratio N:P. A semi-quantitative explanation is sufficient.

2. The ratio N:P of waste released by bodies with increasing N:P is decreasing. The following observations have been made:

N:P Molar Ratio in Body	N/P Ratio in Waste
2	80
3	50
8	20
9	16
20	6

Explain the observed relationship. A semiquantitative explanation is sufficient.

3. Find the size of aquatic bacteria using a substrate with a molar diffusion coefficient of $2 \ 10^{-9} \ m^2/s$. Active bacteria have a metabolism requiring about 12 moles of substrate/m^3s. The substrate concentration in the environment is 0.1 mole/m^3. Which is the size of the freshwater bacteria?

4. If iodine is the limiting factor for brown algae, what is the approximate maximum concentration of brown algae as dry weight per liter in water with a concentration of 0.01 mg iodine/L? Brown algae contain 1.5 g iodine/kg dry matter.

6

The Thermodynamic Interpretation of Ecosystem Growth and Development

---

The best answer raises the most questions.

The growth and development of ecosystems can be described by the three growth forms: growth of biomass, growth of networks, and growth of information. A thermodynamic interpretation of how ecosystems utilize the three growth forms is presented, and it is shown that an increase of eco-exergy and maximum power can be used to describe all three growth forms, while other thermodynamic variables can only be used to describe one or two of the three growth forms. It is clear from observations that ecosystems utilize the free energy to maintain the system far from thermodynamic equilibrium, but when this free energy demand is covered, the ecosystems use the surplus free energy to gain more eco-exergy—meaning to move further away from thermodynamic equilibrium. The consequences for the seasonal changes of ecosystems and for development of new ecosystems are used as illustrative examples.

6.1 Introduction

If the eight conditions formulated in Section 5.6 are realized, it is assumed that carbon-based life and therefore more or less complex ecosystems will emerge. Far from thermodynamic equilibrium systems, ecosystems will, however, decompose and become a dull system without gradients and any structure, if they do not receive an inflow of free energy that can deliver the work energy needed for maintenance of the system far from thermodynamic equilibrium. Ecosystems therefore have to be nonisolated, meaning open to an energy flow. They are actually open, meaning open to matter and energy. This leads to the obvious question: What will happen if the inflow of free energy is bigger than needed to cover the maintenance? This means in Equation (4.24) that

$$d_eEx/dt > d_iEx/dt \qquad (6.1)$$

The seasonal variations of the inflow of free energy by the solar radiation show clearly what is happening when the free energy flow exceeds the need of free energy for maintenance. Ecosystems develop and grow during spring and summer.

The three growth forms are:

1. Growth of biomass
2. Growth of the network
3. Growth of information

They are utilized to store more eco-exergy in the ecosystem by using the surplus free energy that is available when the maintenance is covered. In Part 2, the characteristic properties of ecosystems will be presented. In Chapters 12 and 13, the two characteristic properties—that components of ecosystems cooperate in a network and carry a lot of information—will be discussed. This chapter, however, will explain that ecosystems can move away from thermodynamic equilibrium, meaning they are growing in spite of the constraints of the three thermodynamic laws and of the biochemical constraints. The thermodynamic constraints mean that a considerable amount of free energy is needed to maintain a system far from thermodynamic equilibrium; it would therefore be obvious to ask the question: What will happen if the flow of free energy through the system is more than needed for the maintenance? Can the surplus free energy be utilized? The answer is yes—for growth and development. This and the next chapter look into the details of this question. The answers are formulated as additional thermodynamic propositions that are decisive for the properties of ecosystems, which is the focus of Part 2. As discussed in Section 2.5, the three growth forms are consistent with the widely accepted attributes to describe ecosystem development presented by E.P. Odum. All the attributes can be classified as belonging to one of the three growth forms, and Odum's attributes describe how ecosystems move away from thermodynamic equilibrium. So, what can be denoted the ecological law of thermodynamics (ELT) is valid:

If an (eco)system received more free energy than it needs to cover the maintenance of the system far from thermodynamic equilibrium, then the surplus free energy or exergy is utilized to move further away from thermodynamic equilibrium, meaning that the system gains eco-exergy.

To understand better the thermodynamics behind ecosystem development by the three growth forms, the next two sections will uncover how several key thermodynamic variables change by the three growth forms and how the easily observed seasonal changes of ecosystems can be explained by these thermodynamic variables.

6.2 The Ecosystem Development Described by a Thermodynamic Interpretation of the Three Growth Forms

To illustrate the three growth forms thermodynamically, the following thermodynamic variables will be used:

Eco-exergy (work capacity stored in the ecosystem relative to thermodynamic equilibrium)
Power (throughflow of useful energy = the sum of all free energy flows in the system)
Emergy stored in the ecosystem
Retention time
Exergy consumption, loss, or reduction due to maintenance of the system
Entropy production
Specific exergy of the ecosystem (eco-exergy stored divided by the biomass)
Specific entropy production (entropy production/biomass)

The usual description of ecosystem development illustrated, for instance, by the recovery of Yellowstone National Park after fire, an island born after a volcanic eruption, reclaimed land, etc., is well covered by Odum (1969): At first the biomass increases rapidly, which implies that the percentage of captured incoming solar radiation increases, but also the energy needed for the maintenance. The increase of biomass is mainly in the form of plants that capture the solar radiation almost as an antenna captures radio waves. Growth form I is dominant in this first phase, where exergy stored increases (more biomass, more physical structure to capture more solar radiation), but also the throughflow (of useful energy) increases because the flows are determined by the biomass, and to maintain more biomass also requires more free energy. Furthermore, exergy consumption or loss and the entropy production increase due to increased need of energy for maintenance.

When the percentage of solar radiation captured reaches about 80%, it is not possible to increase the amount of captured solar radiation further (due in principle to the second law of thermodynamics). It is of course completely impossible to exceed 100% but as free energy (eco-exergy) is lost by all energy transformations, about 80% is the upper limit in practice. Therefore, further growth of the physical structure (biomass) cannot improve the energy balance of the ecosystem. In addition, all or almost all the essential elements are in the form of dead or living organic matter, and not as inorganic compounds ready to be used for growth. Growth form I will therefore not proceed, but growth forms II and III can still operate. The ecosystem can still develop and improve the ecological network and can still change r-strategists with K-strategists, small animals and plants with bigger ones, and less developed with more developed organisms with more information genes. A graphic representation of this description of ecosystem development is presented in Figure 6.1, where the solar radiation captured is plotted vs. the eco-exergy stored in the ecosystem in accordance with Table 6.1 As the increase of eco-exergy is proportional to the increase of biomass, the specific eco-exergy will not change in this development phase, and as the free energy used for maintenance with

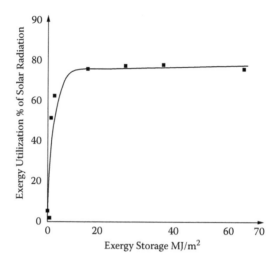

FIGURE 6.1   The exergy captured (taken from Kay and Schneider, 1992; expressed as % of solar radiation) is plotted vs. the exergy stored (unit J/m² or J/m³), calculated from the characteristic compositions of the focal eight ecosystems. The numbers from Table 6.1 are applied to construct this plot. Notice that exergy utilization for maintenance is parallel (proportional) to energy absorbed and the biomass of the ecosystem.

TABLE 6.1   Exergy Utilization and Storage in a Comparative Set of Ecosystems

Ecosystem	Exergy Utilization, %	Exergy Storage, MJ/m²
Quarry	6	0
Desert	2	0.073
Clear-cut forest	49	0.594
Grassland	59	0.940
Fir plantation	70	12.70
Natural forest	71	26.00
Old-growth deciduous forest	72	38.00
Tropical rain forest	70	64.00

good approximation is proportional to the biomass, the specific entropy production is also unchanged.

As will be shown in Chapter 12 in more detail, growth form II—development of the network—implies that the ecosystem obtains a higher efficiency in its application of the incoming free energy. It entails that the ecosystem can move further away from thermodynamic equilibrium. If there is inorganic matter to build up more biomass, growth form I can be enhanced by growth form II, but it is of course also possible that additional organisms that can utilize the various food sources better are contributing to a higher eco-exergy, or organisms with more information are replacing organisms with less information, including that r-strategists are replaced by K-strategists. There are several possibilities that the more effective network is utilized. But they all entail that the

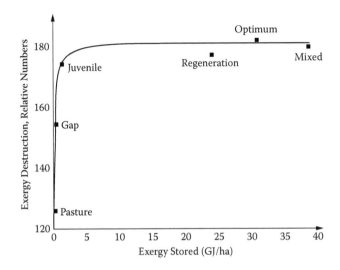

FIGURE 6.2 The plot shows the result by Debeljak (2002). He examined managed and virgin forests in different stages. The gap has no trees, while the virgin forest changes from optimum to mixed to regeneration and back to optimum, although the virgin forest can be destroyed by catastrophic events like fire or storms. The juvenile stage is a development between the gap and the optimum. Pasture is included for comparison.

ecosystem move further away from thermodynamic equilibrium. Most of these possibilities do not entail that the biomass is increasing, or at least it is not increasing proportional to the gain in eco-exergy. This means that the entropy production and the exergy consumption may not increase or increase only slightly, and that the specific exergy and entropy production therefore decrease.

Growth form III—growth of information—means that the β-value in the expression for the calculation of eco-exergy is increasing, and it is tantamount to a gain of the eco-exergy of the ecosystem, and that it is moving further away from thermodynamic equilibrium. As K-strategists often have higher β-values than r-strategists, growth form III often entails a stepwise shift from r-strategists to K-strategists.

Debeljak (2002) has shown that he gets the same shape of the curve when he determines exergy captured and exergy stored in managed and virgin forests at different stages of development (see Figure 6.2). Aquatic ecosystems do not have the same ability as terrestrial ecosystems to capture the sunlight, and different aquatic ecosystems are less similar in their possibilities to capture the sunlight than terrestrial ecosystems. Nevertheless, as shown by Jørgensen (2007), the aquatic ecosystems will also show a Michaelis-Menten-like plot when the exergy storage is plotted vs. the exergy captured, which is almost equal to the exergy used for maintenance, as most captured free energy is used for maintenance. The plot for aquatic ecosystems is shown in Figure 6.3.

Emergy is of course increasing by growth form I, as more solar radiation has been captured and applied to build the additional biomass. As there are several possibilities for the development of ecosystems when a more effective ecological network results, it is necessary to calculate how much emergy will change when the network is getting

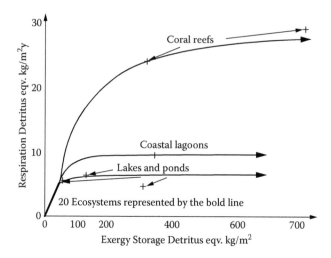

FIGURE 6.3 The eco-exergy storage expressed as kg detritus equivalent/m² is plotted vs. the eco-exergy used for maintenance (respiration) in kg detritus equivalent/m² year for 26 different ecosystems. By multiplication by 18.7 the results are obtained in MJ/m², as the average free energy (eco-exergy) of detritus is 18.7 kJ/g. The respiration is indicated directly in kg/m² year. Three different levels of respiration, namely, for (1) coral reef, (2) fertile lagoons and estuaries, and (3) fertile lakes and ponds, are shown as almost horizontal lines in the figure.

more effective. Organisms that can utilize the various food sources better contribute to a higher eco-exergy and often have a higher sej value, which means that they have more emergy, or organisms with more information are replacing organisms with less information, including that r-strategists are replaced by K-strategists, when the networks get more effective—again this means that emergy is increasing. The organisms late in the food chain usually have higher β-values and are K-strategist opposite organisms early in the food chain, which are usually r-strategists. Therefore, growth form III will almost always entail increased emergy and, of course, also exergy.

The accordance with the eight descriptors listed above includes specific entropy production and specific exergy, and the three growth forms based on this description of ecosystem development are shown in Table 6.2. The table shows clearly that eco-exergy storage and power are the two thermodynamic descriptors that increase with all three growth forms.

Holling (1986), see Figure 6.4, has suggested how ecosystems progress through the sequential phases of renewal (mainly growth form I), exploitation (mainly growth form II), conservation (dominant growth form III), and creative destruction. The latter phase also fits into the three growth forms but will require further explanation. The creative destruction phase is a result of either external or internal factors. In the first case (for instance, hurricanes and volcanic activity), further explanation is not needed, as an ecosystem has to use the growth forms under the prevailing conditions, which are determined by the external factors. If the destructive phase is a result of internal factors, the question is: Why would a system be self-destructive? A possible explanation is that a result of the conservation phase is that almost all nutrients will

TABLE 6.2 Accordance between Growth Forms and Thermodynamic Variables

	Growth Form I	Growth Form II	Growth Form III
Exergy storage	Up	Up	Up
Power/throughflow	Up	Up	Up
Emergy	Up	Equal, maybe up ?	Up
Exergy consumption	Up	Equal	Equal
Retention time	Equal	Up	Up
Entropy production	Up	Equal	Equal
Exergy/biomass = specific exergy	Equal	Up	Up
Entropy/biomass = specific entropy production	Equal	Down	Down

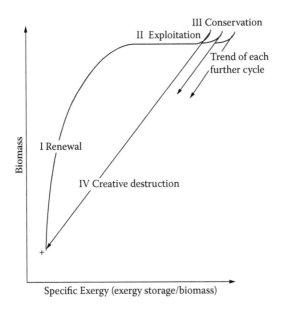

FIGURE 6.4 Holling's four stages are expressed in terms of biomass and specific exergy. Notice that the trend of each future cycle is toward higher exergy storage.

be contained in organisms, which implies that there are no nutrients available to test new and possibly better solutions to move further away from thermodynamic equilibrium or, expressed in Darwinian terms, to increase the probability of survival. This is also implicitly indicated by Holling, as he talks about creative destruction. Therefore, when new solutions are available, it would in the long run be beneficial for the ecosystem to decompose the organic nutrients into inorganic components that can be utilized to test the new solutions. The creative destruction phase can be considered a method to utilize the three other phases and the three growth forms more effectively in the long run.

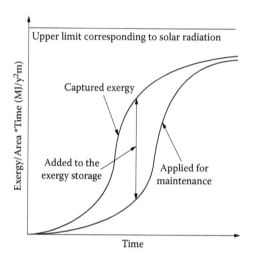

FIGURE 6.5 Exergy utilization of an ecosystem under development is shown vs. time. Notice that the consequence of the growth in exergy is increased utilization of exergy for maintenance.

Most of the free energy of the solar radiation captured by the vegetation is used for maintenance, that is, respiration. The development of the exergy captured and the exergy applied for maintenance may vary considerably from ecosystem to ecosystem, but if we follow E.P. Odum's attributions in our description of ecosystem development, Figure 6.5 gives a typical average picture. The vegetation at an intermediate stage yields the relative biggest contribution to the eco-exergy.

Example 6.1

The values of ecosystem services have been discussed for about the last 10 years. The annual ecosystem services are of course closely related to the annual production. The ecosystem services include all services that an ecosystem offers to the society—not only direct natural resources, but also, for instance, biodiversity, recreational values, cycling of important elements, and purification of air and water. It has therefore been proposed (Jørgensen, 2010) to calculate the annual increase of eco-exergy as a measure of ecosystem services. Eco-exergy is calculated, as has been shown, as the product of biomass and information. It is very important to include the contribution coming from the information, as many of the services are based on the information embodied in the living ecosystem components.

Calculate the annual increase of a number of selected ecosystem types and calculate the approximate corresponding increase of eco-exergy. It is possible to also calculate the money values of the annual eco-exergy by using the electricity cost, about 1.4 U.S. cents/MJ.

Solution

The table below presents the results of the calculations for a number of ecosystems. The annual production of the ecosystems can be found on the Internet.

Annual Work Capacity Increase by Various Ecosystems

Ecosystem Type	MJ/m² Year Biomass	β-Value (average)	Eco-Exergy (GJ/ha year)
Desert	0.9	230	2,070
Open sea	3.5	68	2,380
Coastal zones	7.0	69	4,830
Coral reefs, estuaries	80	120	96,000/93,500
Lakes	11	85	9,350
Coniferous forests	15.4	350	53,900
Deciduous forests	26.4	380	100,000
Temperate rain forests	39.6	380	150,000
Tropical rain forests	80	370	300,000
Tundra	2.6	280	7,280
Croplands	20.0	210	42,000
Grassland	7.2	250	18,000
Wetlands	18	250	45,000

6.3 Seasonal Changes

Exergy storage and utilization patterns follow the seasonal trends in biomass, through-flow (power), and informational characteristics. In winter, biomass and information content are at seasonal lows. In spring, the flush of new growth (dominantly form I) produces rather quickly a significant biomass component of exergy (Figure 6.6), but the information component remains low due to the fact that most active flora, fauna, and microbiota of this nascent period tend to be lower phylogenetic forms. These lower forms rapidly develop biomass but make relatively low informational contributions to the stored exergy.

Figure 6.7 shows a Danish beech forest on May 6. From about April 20 to about May 1 the beech forest in Denmark bursts into leaves and the biomass increases very rapidly during a few days. Growth form I is dominant at this time of the year. As the growing season advances, in summer, growth forms II and III become successively dominant. Following the expansion of system organization that this represents, involving proliferation of food webs and interactive networks of all kinds, and all that this implies, waves of progressively more advanced taxonomic forms can now be supported to pass through their phenological and life cycles. Albedo and reflection are reduced, dissipation increases to seasonal maxima following developing biomass, and as seasonal maxima are reached, further increments taper to negligible amounts (Figure 6.6). The biotic production of advancing summer reflects more and more advanced systemic organization, manifested as increasing accumulations of both biomass and information to the exergy stores. In autumn, the whole system begins to unravel and shut down in preadaptation to winter, the phenological equivalent of senescence. Networks shrink, and with

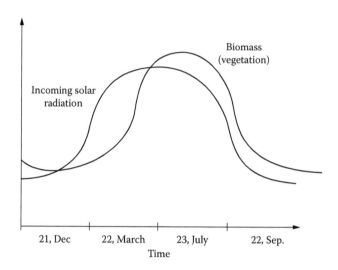

FIGURE 6.6 The seasonal changes in incoming solar radiation and biomass (vegetation) are shown for a typical temperate ecosystem. The x-axis gives the time of the year, while the y-axis shows either the free energy of the incoming solar radiation or the biomass per unit of area or volume. The slope of the curve for biomass indicates the increase in eco-exergy due to growth form I. Growth form I can continue as long as the captured solar radiation is bigger than the eco-exergy applied for maintenance, or as long as there is enough inorganic matter to build new biomass. Therefore, the biomass often has its maximum at about August 1. The biomass is at minimum around the first of February, because at that time the captured exergy and the exergy applied for maintenance (although low) is in balance.

this all attributes of exergy storage, throughflow, and information transfer decline as the system slowly degrades to its winter condition. Biological activity is returned mainly to the more primitive life forms as the ecosystem itself returns to more primitive states of exergy organization required for adaptation to winter. The suggestion from phenology is that the exergetic principles of organization apply also to the seasonal dynamics of ecosystems.

> Notice that the eco-exergy continues to increase until about August 1, because the free energy captured by the vegetation is still bigger than the free energy needed for maintenance. It shows the ability of the ecosystems to capture the extra free energy and store it as biomass, structure, and information.

6.4 New Ecosystems

It would be interesting to follow the development of eco-exergy when new ecosystems are formed. If the eco-exergy is actually increasing for new ecosystems, it would be a strong support for the proposition presented in Section 6.1 about the use of the surplus free energy to move further away from thermodynamic equilibrium. As illustration for the development of eco-exergy for new ecosystems we have chosen Surtsey, a volcanic

FIGURE 6.7 (See color insert.) Danish beech forest, May 6. During a period of about 10 days the forest bursts into leaves. The biomass increases very rapidly. Growth form I is dominant.

island south of Iceland. The island is 150 ha and was formed in 1962 by a volcanic eruption, and the ecological development on the island has been followed by Reykjavik University since 1964. The first year and a half it was impossible to start investigation on the island due to the heat. The development of eco-exergy of plants and birds is shown in Figure 6.8. As seen the eco-exergy has increased since 1964, and up to the year 2000 (indicated on the graph as 100) approximately exponentially, which is supported by a log graph in Figure 6.9. It is expected that the eco-exergy is growing as a first-order reaction, and therefore exponentially, because the growth is proportional to the growth by the three growth forms already obtained.

Examinations of the development of other new ecosystems have shown similar trends, and although the observations for the development of Surtsey were particularly dense and detailed, it seems possible to conclude that new ecosystems will with good approximation show an exponential increase of eco-exergy, completely in accordance with the proposition presented in Section 6.1.

6.5 Summary of Important Points in Chapter 6

1. When an open system receives a higher free energy flow than needed to cover the maintenance of the system, the additional free energy will be used to move the system further away from thermodynamic equilibrium—to gain eco-exergy.
2. Ecosystems have a particular ability to utilize the additional free energy to move away from thermodynamic equilibrium, namely, by growth of biomass, the ecological network, and information.

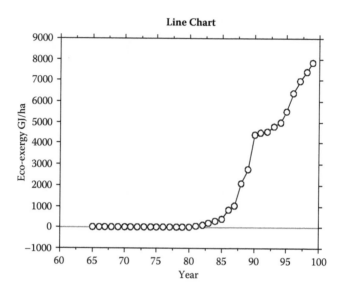

FIGURE 6.8 (See color insert.) The development of eco-exergy on the island of Surtsey, a volcanic island formed south of Iceland by a volcanic eruption in 1962. The island is 150 ha and the eco-exergy for plants and birds has been found in GJ/ha from 1964 to 2000 (indicated as 100 on the graph). The eco-exergy is approximately increased exponentially; see also Figure 6.9.

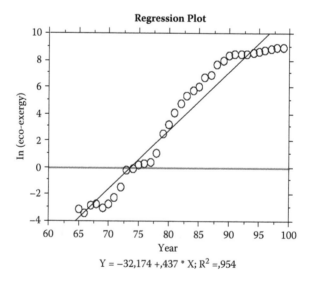

$$Y = -32{,}174 + {,}437 * X; R^2 = {,}954$$

FIGURE 6.9 (See color insert.) Regression plot of log eco-exergy in GJ/ha (see Figure 6.8) vs. the years 1964–2000 (indicated on the graph as 100). As the regression plot with good approximation is linear with $R^2 = 0.954$, it is possible to conclude the eco-exergy on the island with good approximations has increased exponentially.

3. The ecosystems usually utilize the flow of free energy (solar radiation) by growth of biomass, which will increase the efficiency by which the solar radiation is captured by the ecosystem up to about the possible maximum of 75–80%, which is physically possible.

4. When growth form I (growth of biomass) slows down because the efficiency of capturing the solar radiation is approaching 75–80%, or because there is only very little inorganic matter to build biomass, the two other growth forms—growth of the network and information—are gradually taken over.

5. The graph free energy captured, biomass, or free energy used for maintenance vs. eco-exergy stored in the ecosystem has a Michaelis-Menten-like shape.

6. Points (1) and (2) are able to explain the seasonal variations of free energy captured by ecosystems and the eco-exergy stored in ecosystems as a function of the season.

Exercises/Problems

1. Explain why is it an effective strategy for ecosystems to utilize the surplus free energy of solar radiation in the firsthand for growth of biomass?

2. Explain why growth form I (growth of biomass) has clearer limitations than the two other growth forms.

3. Explain why a Michaelis-Menten-like graph, as shown in Figures 6.1–6.3, results from the succession of the three growth forms.

4. What would be the fate of ecosystems that have the curve applied for maintenance above the curve of captured eco-exergy?

5. An ecosystem at the early stage (for instance, an agriculture field in the early spring or a gap in a forest) has the following distribution of the inflowing free energy (solar radiation): 30% is used for maintenance (respiration) and evapotranspiration, 20% is used for growth (gain of eco-exergy), and 50% is reflected. Indicate how this free energy distribution is changed as the ecosystem develops.

6. Make a graph similar to Figure 6.6 for the diurnal variations.

7

The Ecological Law
of Thermodynamics

We cannot dispose the future.
But we can propose the trends.

The ecological law of thermodynamics (ELT) is presented with additions: A system that receives a throughflow of exergy (high-quality energy) will try to utilize the exergy flow to move away from thermodynamic equilibrium, and if more combinations of components and processes are offered to utilize the exergy flow, the system will select the organization that gives the system as much exergy content (storage) as possible, i.e., maximizes dEx/dt. Several ecological observations and rules that support ELT will be presented, including the application of structurally dynamic models (SDMs). Also, evolutionary theories comply with ELT.

7.1 Introduction: Darwin's Theory

Ecosystems are open systems and receive free energy from solar radiation. As discussed in Chapter 6, the free energy is used first to cover the free energy needed for maintaining a system very far from thermodynamic equilibrium. If there is more free energy available than needed for the maintenance, the surplus free energy is used to move the system further away from thermodynamic equilibrium by use of the three growth forms. There are, however, many possible pathways to move away from thermodynamic equilibrium, because ecosystems have many components and are linked in a complex ecological network. This implies that the three growth forms can be used by the ecosystem to move away from thermodynamic equilibrium in an almost astronomically high number of possibilities. So, which one of these numerous possibilities will be selected?

Darwin has given the answer: survival of the fittest, meaning that the species that have properties that are best fitted to the prevailing conditions, determined by the forcing functions or constraints, will survive and grow. Survival is represented by biomass and information. Survival is in this context a question about the survival of the genes—the information is important because the survival of the genes ensures that the processes, the biomass, and the functions of the ecosystem are conserved for the next generation of organisms, and therefore on a longer-term basis. The more biomass, the more difficult it will be to eliminate the biomass (Svirezhev, 1990), and more information means that the resources (necessary element to build the biomass and the use of the solar radiation) are used more effectively. Biomass and information are expressed by eco-exergy, and as all the species in an ecosystem are interrelated and dependent on each other through a complex ecological network, the eco-exergy of the ecosystem can be used to account for the survival of the entire ecosystem. These holistic considerations are included in the translation of Darwin's theory to thermodynamics by the ecological law of thermodynamics (ELT), which is one of the main focuses of this chapter. Darwin's theory focuses on the survival of the organisms, but all the organisms are dependent on all other organisms because they are linked in a network and affect and influence each other. It is therefore absolutely necessary to consider the survival of the entire ecosystem with all the living and interdependent components. To summarize, ELT is a translation of Darwin's theory to thermodynamics and at the same time an expansion from the organisms—or according to the neo-Darwinian interpretation, the genes of the organisms—to the entire ecosystem.

Without the genes, or rather a heritage system, there would have been no development or evolution. The interplay between the three growth forms also plays an important role, because the growth of biomass and networks may create constraints that inevitably will influence the third growth form, the information, that is embodied in the genes. The neo-Darwinian theory has taught us that the adaptation occurs through natural selection of changed genetic variations. It is a part of the story, but not the full story. Let us denote it heredity system number one. It includes the so-called Hox genes, which play a pivotal role in specifying regional identity in body plans. It has been suggested that increasing complexity of body plans during evolution might be causally correlated with increasing complexity of the Hox gene complexes (Mayr, 2001).

It has been shown that cells can transmit information to daughter cells through non-DNA (epigenetic) inheritance. Let us call it heredity system number two. In addition, many animals transmit to others by behavioral means, which may be considered a third heredity system. For instance, the bear mother shows the cubs by her behavior which food it is best to eat and how to catch salmons. Finally, we have the symbol-based heredity system, particularly language, which has played a more and more important role through evolution as a heredity system. The bear mother can also by sounds warn the cubs about dangers. Language plays an enormous role for *Homo sapiens* as a heredity system. The presence of the four heredity systems is able to explain why we have observed at least in some phases a surprisingly rapid evolution. When all four inheritance systems and the interactions between them are taken into account, we get a different view of Darwinian evolution (Jablonka and Lamb, 2006). The possibilities for development and evolution are under all circumstances more complex than we thought a few decades ago.

Darwin's theory builds on four principles or properties:

1. **Reproduction.** The organisms are able to reproduce, and every population has such high fertility that its size would increase exponentially if not constrained.

2. **Inheritance.** The properties of the offspring are almost unchanged when inherited from the parents or from the mother cell. This is the basis of heredity system number one: the genes. Evolution would be impossible without the possibilities of building on the shoulders of what has already been achieved. The change of the genes may be random—by mutation and sexual recombinations—but once a good gene has been found, i.e., genes that give a high probability of survival and further growth, the information will be transferred to the next generation due to the genes.

3. **Variation.** Not all organisms are identical, but they are all more or less different. The variation is needed to have a selection. If all the organisms were identical, there would be no differences in survival and growth, and therefore no selection. The genes are different from organism to organism even among organisms of the same species. So, the selection takes place among the differences in properties of the phenotypes, resulting in currently changing genomes generation after generation. A selection due, for instance, to the introduction of new constraints will, on the firsthand, imply that the distribution in the variation of a property will move in the direction of what yields the best fitness (see Figure 7.1).

4. **Competition.** The resources are limited. Therefore, only some of the organisms can survive. The other will be outcompeted or eliminated in the fight about the resources. The heritable variation affects the success of the organisms in surviving and multiplying.

The variation, principle 3, represents chance in the Monod sense: Evolution is a trade-off between chance and necessity (Monod, 1972), and the selection, reproduction, and heritage (principles 1 and 2), as a result of limited resources (principle 4), represent the necessity.

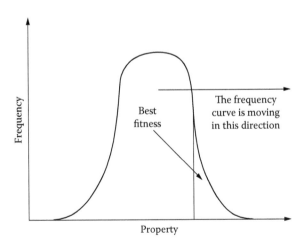

FIGURE 7.1 A selection due, for instance, to the introduction of new constraints will in the firsthand imply that the distribution in the variation of a property will move in the direction of what yields the best fitness.

7.2 The Ecological Law of Thermodynamics (ELT)

The tentative ecological law of thermodynamics (ELT) gives information on which of the many possible processes will be realized as a result of a competition among more possibilities than the flow of exergy can accomplish. Ulanowicz (1997) has introduced the expression "the propensity of ecosystems" to stress that ecosystems and their forcing functions encompass many random components and have an enormous complexity, which makes accurate predictions impossible. Propensities are weighted or conditional probabilities that are inherent features of changing situations or occasions rather than absolute properties of relational processes and components. Ever since quantum mechanics introduced indeterminacy, it has become increasingly easy to recognize that we do actually live in a world of propensities, with an unfolding process of realizing possibilities and creating new possibilities. It seems therefore advantageous to include this expression in the formulation of the pattern of ecosystem theories, and also in the formulation of the tentative fourth law of thermodynamics, or the ecological law (hypothesis) of thermodynamics.

The tentative law, according to the latest formulation, asserts that:

> A system that receives a throughflow of exergy (high-quality energy) will try to utilize the exergy flow to move away from thermodynamic equilibrium, and if more combinations of components and processes are offered to utilize the exergy flow, the system will select the organization that gives the system as much exergy content (storage) as possible, i.e., maximizes dEx/dt.

A few very competent ecologists have expressed preference for a formulation where the flow of eco-exergy is replaced by a flow of free energy, which of course is fully acceptable and makes the formulation closer to classic thermodynamics. However, eco-exergy cannot be replaced by free energy because it is a free energy *difference* between the system and the same system at thermodynamic equilibrium. The reference state is therefore different from ecosystem to ecosystem, which is considered in the definition of eco-exergy. In addition, free energy and eco-exergy are not state functions far from thermodynamic equilibrium—just consider the immediate loss of eco-exergy when an organism dies (compare with Section 4.9). Before the death the organism has high eco-exergy because it can utilize the enormous information that is embodied in the amino acid sequence of the enzymes that are controlling the life processes. At death the organism loses immediately the ability to use this information, which therefore becomes worthless.

The support for the validity of the tentative law in its present formulation is strong and may be summarized in the following points:

1. It may be considered a translation of Darwin's theory to thermodynamics and is consistent with the basic thermodynamic laws. The selected organization is the one that offers the most survival that can be measured as exergy. The selection is in accordance with the latest formulations of Darwin's theory still taking place on the levels of species. The species are surviving, growing, and fighting for the resources. All the species are, however, connected in an ecological, cooperative, synergistic network and are dependent on each other. The survival is under the

prevailing conditions, which include the presence of all the components in the ecological network. All the species in the ecological network influence all the other species. The result is therefore that the entire ecological network gets as much survival and therefore eco-exergy as possible under the prevailing conditions.

2. The application of the hypothetical law in models gives (many) results that are consistent with ecological observations; see Jørgensen (2002) and Jørgensen and Svirezhev (2004). It includes the use of structurally dynamic models (SDMs).

3. Many ecological observations, including our description (image) of the evolution, can be explained by the presented hypothesis (see Jørgensen, 2002; Jørgensen et al., 2000; Jørgensen and Svirezhev, 2004; Jørgensen et al., 2007).

4. Proteins carry out all the biochemical reactions in the cells and their physico-chemical properties are the prerequisite for the existence of cells. Their amino acid sequence, which is fundamental for enzymatic properties, is coded by the information embodied in the genome. Life is information; see also the further discussion of this topic in Chapter 13.

The next section presents a few case studies from Jørgensen (1997, 2002) and Jørgensen et al. (2000) supporting the presented eco-exergy storage hypothesis (ELT), but the maximum power principle could also have been applied. Section 7.4 is devoted to structurally dynamic models (SDMs), which are models that describe the changes in structures of ecosystems due to adaptation and changes in species composition by using ELT. As SDMs in 23 cases have been able to predict the actual structural changes observed, this success of the application of SDMs can be considered a strong support for ELT. Section 7.5 uncovers the compliance between the evolutionary theories and ELT, which of course is an additional support for ELT.

7.3 Some Basic Ecological Observations (Rules) That Can Be Explained by ELT

Below are presented a few case studies from Jørgensen (1997, 2002) and Jørgensen et al. (2000) supporting ELT. More examples can be found in these references and in Jørgensen et al. (2007).

1. **Size of genomes.** In general, biological evolution has been toward organisms with an increasing number of genes (Futuyma, 1986). If a direct correspondence between free energy and genome size (compare with Figure 4.7) is assumed, this can reasonably be taken to reflect increasing eco-exergy storage accompanying the increased information content and processing of higher organisms. This is discussed in more detail in the evolutionary theories context (see Section 7.5).

2. **Le Chatelier's principle.** The exergy storage hypothesis might be taken as a generalized version of Le Chatelier's principle. Biomass synthesis can be expressed as a chemical reaction:

energy + nutrients = molecules with more free energy (exergy) and
organization + dissipated energy

TABLE 7.1 Yields of kJ and ATPs per Mole of Electrons, Corresponding to 0.25 Moles of CH_2O

Oxidized (Carbohydrates) Reaction	kJ/mole e–	ATPs/mole e–
$CH_2O + O_2 = CO_2 + H_2O$	125	2.98
$CH_2O + 0.8 NO_{3-} + 0.8 H^+ = CO_2 + 0.4 N_2 + 1.4 H_2$	119	2.83
$CH_2O + 2 MnO_2 + H^+ = CO_2 + 2 Mn^{2+} + 3 H_2O$	85	2.02
$CH_2O + 4 FeOOH + 8 H^+ = CO_2 + 7 H_2O + Fe^{2+}$	27	0.64
$CH_2O + 0.5 SO_4^{2-} + 0.5 H^+ = CO_2 + 0.5 HS^- + H_2O$	26	0.62
$CH_2O + 0.5 CO_2 = CO_2 + 0.5 CH_4$	23	0.55

Note: The released energy is available to build ATP for various oxidation processes of organic matter at pH = 7.0 and 25°C.

According to Le Chatelier's principle, if energy is put into a reaction system at equilibrium the system will shift its equilibrium composition in a way to counteract the change. This means that more molecules with more free energy and organization will be formed. If more pathways are offered, those giving the most relief from the disturbance (displacement from equilibrium) by using the most energy, and forming the most molecules with the most free energy, will be the ones followed in restoring equilibrium.

3. **Sequence of organic matter oxidation.** The sequence of biological organic matter oxidation (e.g., Schlesinger, 1997) takes place in the following order: by oxygen, by nitrate, by manganese dioxide, by iron (III), by soleplate, and by carbon dioxide. This means that oxygen, if present, will always outcompete nitrate, which will outcompete manganese dioxide, and so on. The amount of exergy stored as a result of an oxidation process is measured by the available kJ/mole of electrons, which determines the number of adenosine triphosphate molecules (ATPs) formed. ATP represents exergy storage of 42 kJ/mole. Usable energy as exergy in ATPs decreases in the same sequence as indicated above. This is expected if the exergy storage hypothesis is valid (Table 7.1). If more oxidizing agents are offered to a system, the one giving the highest storage of free energy will be selected.

In Table 7.1, the first (aerobic) reaction will always outcompete the others because it gives the highest yield of stored eco-exergy. The last (anaerobic) reaction produces methane; this is a less complete oxidation than the first because methane has greater exergy content than water.

Example 7.1

Explain:
1. Why denitrification requires anaerobic conditions.
2. Why phosphorus is released from the sediment in aquatic ecosystems with thermocline.

Solution

1. Nitrate can only be used for oxidation when oxygen is not present.
2. The bottom water in aquatic ecosystems with thermocline has a low redox potential because the sediment is usually rich in organic matter and oxygen cannot be transferred to the bottom water due to the thermocline. This implies that iron is in the form of Fe(II), and iron(II) phosphate is considerably more soluble than iron(III) phosphate.

4. **Formation of organic matter in the primeval atmosphere.** Numerous experiments have been performed to imitate the formation of organic matter in the primeval atmosphere on earth 4 billion years ago (Morowitz, 1968). Energy from various sources was sent through a gas mixture of carbon dioxide, ammonia, and methane. There are obviously many pathways to utilize the energy sent through simple gas mixtures, but mainly those forming compounds with rather large free energies (amino acids and RNA-like molecules with high exergy storage, decomposed when the compounds are oxidized again to carbon dioxide, ammonia, and methane) will form an appreciable part of the mixture (according to Morowitz, 1968). See Figure 4.1.

5. **Photosynthesis.** There are three biochemical pathways for photosynthesis: (1) the C3 or Calvin-Benson cycle, (2) the C4 pathway, and (3) the crassulacean acid metabolism (CAM) pathway. The latter is least efficient in terms of the amount of plant biomass formed per unit of energy received. Plants using the CAM pathway are, however, able to survive in harsh, arid environments that would be inhospitable to C3 and C4 plants. CAM photosynthesis will generally switch to C3 as soon as sufficient water becomes available (Shugart, 1998). The CAM pathways yield the highest biomass production, reflecting exergy storage, under arid conditions, while the other two give highest net production (exergy storage) under other conditions. Generally, the biomass production is selected in a direction that is consistent, under the prevailing conditions, with the exergy storage hypothesis (ELT).

6. **Leaf size.** Givnish and Vermelj (1976) observed that leaves optimize their size (thus mass) for the conditions. This may be interpreted as meaning that they maximize their free energy content. The larger the leaves, the higher their respiration and evapotranspiration, but the more solar radiation they can capture. Deciduous forests in moist climates have a leaf area index (LAI) of about 6 (see also Example 3.1). Such an index can be predicted from the hypothesis of highest possible leaf size, resulting from the trade-off between having leaves of a given size vs. maintaining leaves of a given size (Givnish and Vermelj, 1976). The size of leaves in a given environment depends on the solar radiation and humidity regime, and while, for example, sun and shade leaves on the same plant would not have equal exergy contents, in a general way leaf size and LAI relationships are consistent with the hypothesis of maximum exergy storage.

7. **Biomass packing.** The general relationship between animal body weight, W, and population density, D, is D = A/W, where A is a constant (Peters, 1983). Highest packing of biomass depends only on the aggregate mass, not the size of individual organisms. This means that it is biomass rather than population size that is maximized in an ecosystem, as density (number per unit area) is inversely proportional to the weight of the organisms. Of course, the relationship is complex. A given mass of mice would not contain the same exergy or number of individuals as an equivalent weight of elephants. Also, genome differences (Example 7.1) and other factors would figure in. Later we will discuss exergy dissipation as an alternative objective function proposed for thermodynamic systems. If this were maximized rather than storage, then biomass packing would follow the relationship D = A/W 0.65–0.75 (Peters, 1983). As this is not the case, biomass packing and the free energy associated with this lend general support for the exergy storage hypothesis.

8. **Cycling.** If a resource (for instance, a limiting nutrient for plant growth) is abundant, it will typically recycle faster. This is considered strange, because recycling is not needed when a resource is nonlimiting. A modeling study (Jørgensen, 1997) indicates that free energy storage increases when an abundant resource recycles faster. The result is shown in Figure 7.2. The ratio, R, of nitrogen (N) to phosphorus (P) cycling that gives the highest exergy is plotted in a logarithmic scale vs. log (N/P). The plot in Figure 7.2 is also consistent with empirical results (Vollenweider, 1975). Of course, one cannot "inductively test" anything with a model, but the indications and correspondence with data do tend to support in a general way the exergy storage hypothesis. The cycling ratio giving the highest ascendency is also correlated similarly to the N/P ratio (personal communication with R. Ulanowicz). In light of the close relationship between exergy and ascendency, this result is not surprising (see above, Jørgensen, 1995a; Ulanowicz, 1997).

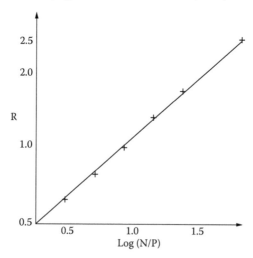

FIGURE 7.2 Log-log plot of the ratio of nitrogen to phosphorus turnover rates, R, at maximum exergy vs. the logarithm of the nitrogen/phosphorus ratio, log N/P. The plot is consistent with Vollenweider (1975).

9. **Foraging behavior of European starlings.** During the spring and summer seasons, starlings typically forage on lawns or pastures for leatherjackets (larvae of tipulid flies). When they are gathering food for the young, they hold captured leatherjackets at the base of the bill. The more leatherjackets a starling has in its bill, the more difficult it is to capture the next one. Therefore, the time between captures increases as more prey is caught. The rate at which the parent delivers food to its young is the number of prey caught divided by the time used for the foraging trip, which includes the time spent on the site and the time spent on traveling between the site and the nest. It has been shown that a starling is able to maximize the rate at which food is delivered (Ricklefs, 2000). A starling is able to find the optimum between the number of prey, the time spent to capture the prey, and the time needed for flying between site and nest. As a result, the starling family has a maximum increase of eco-exergy. The optimum foraging theory or hypothesis is further discussed in Chapter 15.

Example 7.2

Certain parasites are able to attack and damage one ear of insects, but never attack the second ear. How can these observations be explained by ELT?

Solution

If the parasites would damage both ears, the insects would be very easy victims for bats, because the insects could not hear them. It is therefore not beneficial for the parasites to attack and damage two ears, because it would imply that their hosts certainly would be eaten by bats. The result is that the parasites and the insects survive and therefore contribute more to the eco-exergy of the system.

7.4 Structurally Dynamic Models (SDMs)

If we follow the generally applied modeling procedure presented in most textbooks on ecological modeling, we will develop a model that describes the processes in the focal ecosystem, but the parameters will represent the properties of the state variables as they are in the ecosystem during the examination period. They are not necessarily valid for another period because we know that an ecosystem can regulate, modify, and change them, if needed as a response to changes in the existing conditions (see Figure 7.3), determined by the forcing functions and the interrelations between the state variables. Our present models have rigid structures and a fixed set of parameters, meaning that no changes or replacements of the components are possible. However, we need to introduce parameters (properties) that can change according to variable forcing functions and general conditions. The state variables (components) optimize continuously the ability of the system to move away from thermodynamic equilibrium. The state variables can change by adaptation or by change of the species composition. There are always several

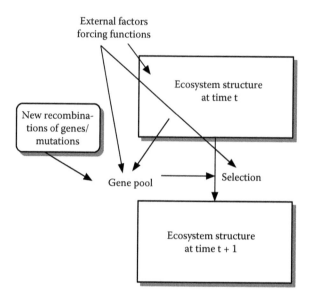

FIGURE 7.3 Conceptualization of how the external factors steadily change the species composition. The possible shifts in species composition are determined by the gene pool, which is steadily changed due to mutations and new sexual recombinations of genes. The development is, however, more complex; see also Section 7.1. This is indicated by (1) arrows from "external factors" to "gene pool" and "selection" that forcing functions can change the gene pool and the selection pressure; and (2) an arrow from "structure" to "gene pool" to account for the possibilities that the species can, to a certain extent, change their own gene pool.

species waiting in the wings ready to take over, if they are better survivors. The idea is therefore to test if a change of the most crucial parameters would be able to move the system more away from thermodynamic equilibrium, and if that is the case, to use that set of parameters, because it expresses the adaptation and the shifts in species composition that take place.

The model type that can account for the change in species composition as well as for the ability of the species, i.e., the biological components of our models, to change their properties, i.e., to adapt to the existing conditions imposed on the species, is already denoted in the structurally dynamic model, to underline that it is able to capture structural changes. Such models may also be called the next or fifth generation of ecological models to indicate that they are radically different from previous modeling approaches and can do more, namely, describe adaptations and shifts in species composition (see also Jørgensen 1992a, 1992b, 1994a, 1994b, 1995a, 1995b).

It could be argued that the ability of ecosystems to replace present species with other, better-fitted species could be considered by constructing models that encompass actual species for the entire period that the model attempts to cover. This approach, however, has two essential disadvantages. The model becomes, first of all, very complex, as it will contain many state variables for each trophic level. Therefore, the model will contain many more parameters that have to be calibrated, and this will introduce a high uncertainty to the model and will render the application of the model very case specific (Nielsen, 1992a, 1992b). In addition, the model will still be rigid and not allow the model to have continuously changing parameters due to adaptation, even without changing the species composition.

Straskraba (1979) uses a maximization of biomass to express the distance from thermodynamic equilibrium. The model computes the biomass and adjusts one or more selected parameters to achieve the maximum biomass at every instance. The model has a routine that computes the biomass for all possible combinations of parameters within a given realistic range. The combination that gives the maximum biomass is selected for the next time step, and so on. Biomass, however, can only be used when only one state variable is adapting or shifted to other species.

Eco-exergy (work capacity) calculated for ecosystems by the use of a special reference system, namely, the same system but at thermodynamic equilibrium at the same temperature and pressure, has been used widely as a goal function in ecological models for development of SDMs. Two of the available and most illustrative case studies will be presented and discussed below. Eco-exergy has two pronounced advantages as goal function. It is defined far from thermodynamic equilibrium, and it is related to the state variables, which are easily determined, modeled, or measured, opposite, for instance, maximum power that is related to the flows, which are more rarely determined and more difficult to measure. Furthermore, eco-exergy can also be applied when two or more species are adapting or shifted to other species, which is often the case. Phytoplankton and zooplankton are, for instance, often changed simultaneously when the forcing function for lakes is changed.

The thermodynamic variable eco-exergy has been applied to develop structurally dynamic models in 23 cases; see Zhang et al. (2010). The 23 case studies are:

1-8. Eight eutrophication models of six different lakes
9. A model to explain the success and failure of biomanipulation based on removal of planktivorous fish
10. A model to explain under which circumstances submerged vegetation and phytoplankton are dominant in shallow lakes
11. A model of Lake Balaton that was used to support the intermediate disturbance hypothesis
12-15. Small population dynamic models
16. Eutrophication model of the Lagoon of Venice
17. Eutrophication model of the Mondego Estuary
18. Evolution model of Darwin's finches
19. An ecotoxicological model focusing on the influence of copper on zooplankton growth rates
20. A model of the interaction between parasites and birds
21. The structurally dynamic model, including Pamolare 1 applied on Lake Fure in Denmark
22. Lake Chazas in Spain (Marchi et al., 2011)
23. An individual-based model (IBM) to show that conjugation is able to provide a better combination of parameters to obtain a higher eco-exergy level

Below are given two illustrative examples of structural dynamic models, namely, 18 and 19 from the above list of case studies, because they are both relatively simple, easily understandable models, where the structural changes have been observed and are easy to conceive.

Figure 7.4 shows how SDMs are developed by determining the change of the parameters that are needed to describe adaptation and shifts in species composition by using eco-exergy as the goal function. Figure 7.5 illustrates the theoretical considerations behind the use of SDM.

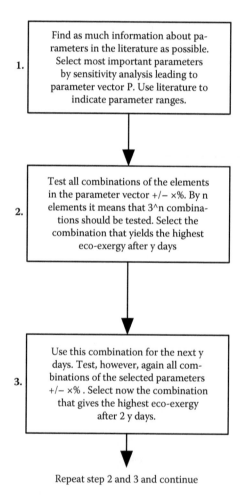

1. Find as much information about parameters in the literature as possible. Select most important parameters by sensitivity analysis leading to parameter vector P. Use literature to indicate parameter ranges.

2. Test all combinations of the elements in the parameter vector +/− x%. By n elements it means that 3^n combinations should be tested. Select the combination that yields the highest eco-exergy after y days

3. Use this combination for the next y days. Test, however, again all combinations of the selected parameters +/− x% . Select now the combination that gives the highest eco-exergy after 2 y days.

Repeat step 2 and 3 and continue

FIGURE 7.4 The procedure used for the development of structurally dynamic models is shown. SDMs have been developed successfully in 23 cases. The observed structural changes have been predicted by the model with an acceptable standard deviation that is general for ecological modeling. The state variables are in addition generally predicted with a smaller standard deviation than for the same ecological models not considering structural changes.

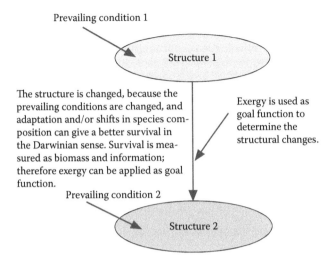

FIGURE 7.5 (See color insert.) The theoretical considerations behind the application of SDM.

7.4.1 Development of SDM for Darwin's Finches

The development of a structurally dynamic model for Darwin's finches illustrates the advantages of SDMs very clearly (see details in Jørgensen and Fath, 2004). The model reflects—as all models—the available knowledge, which in this case is comprehensive and sufficient to validate even the ability of the model to describe the changes in the beak size as a result of climatic changes, that are causing changes in the amount, availability, and quality of the seeds that make up the main food item for the finches. The medium ground finches, *Geospiza fortis*, on the island Daphne Major were selected for these modeling cases due to very detailed case-specific information found in Grant (1986). The model has three state variables: seed, Darwin's finches adult, and Darwin's finches juvenile. The juvenile finches are promoted to adult finches 120 days after birth. The mortality of the adult finches is expressed as a normal mortality rate plus an additional mortality rate due to food shortage and an additional mortality rate caused by a disagreement between bill depth and the size and hardness of seeds. Due to a particular low precipitation in 1977–1979 the population of the medium ground finches declined significantly, and the beak size increased at the same time about 6%. An SDM was developed to be able to describe this adaptation of the beak size due to bigger and harder seeds as a result of the low precipitation.

The beak depth can vary between 3.5 and 10.3 cm according to Grant. The adapted beak size is furthermore equal to the square root of D*H, where D is the diameter and H the hardness of the seeds. Both D and H are dependent on the precipitation, particularly from January to April. The coordination or fitness of the beak size with D and H is a survival factor for the finches. The fitness function is based on the seed handling time, and it influences the mortality as mentioned above, but also has an impact on the number of eggs laid and the mortality of the juveniles. The growth rate and mortality rate of the seeds are dependent on the precipitation and the temperature, which are forcing functions known as f(time). The food shortage is calculated from the food required by the finches, which is known according to Grant and the actual available food according to the state function seed. How the food shortage influences the mortality of the adults and juveniles can be found in Grant (1986). The seed biomass and the number of finches are known as a function of time for the period 1975–1982 (see Grant, 1986). The observations of the state variables from 1975 to 1977 were applied for calibration of the model, focusing on the following parameters:

1. The influence of the fitness function on (a) the mortality of adult finches, (b) the mortality of juvenile finches, and (c) the number of eggs laid.
2. The influence of food shortage on the mortality of adult and juvenile finches is known (Grant, 1986). The influence is therefore calibrated within a narrow range of values.
3. The influence of precipitation on the seed biomass (growth and mortality).

All other parameters are known from the literature (see Grant, 1986).

The eco-exergy density is calculated (estimated) as 275 × the concentration of seed + 980 × the concentration of the finches (see Table 4.1). Every 15 days it is determined whether a feasible change in the beak size (the generation time and the variations in the beak size are taken into consideration) will give a higher eco-exergy. If this is the case, then the beak size is changed accordingly. The modeled changes in the beak size were confirmed by the observations. The model results of the number of Darwin's finches are compared with the observations in Figure 7.6. The standard deviation between modeled and observed values was 11.6%. For the validation, the correlation coefficient, r^2, for modeled vs. observed values is 0.977. The results of a nonstructural dynamic model

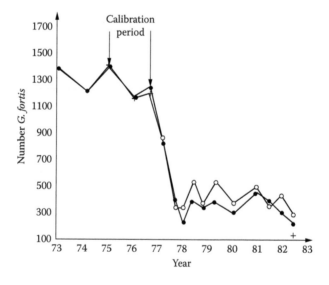

FIGURE 7.6 The observed number of finches (●) from 1973 to 1983, compared with the simulated result (○). 75 and 76 were used for calibration and 77/78 for the validation.

would not be able to predict the changes in the beak size, and would therefore give too low values for the number of Darwin's finches because their beak would not adapt to the lower precipitation, yielding harder and bigger seeds. The calibrated model not using the eco-exergy optimization for the structurally dynamic models in the validation period 1977–1982 resulted in complete extinction of the finches. A nonstructurally dynamic model—a normal biogeochemical model—could therefore not describe the impact of the low precipitation, while the SDM gave an approximately correct number of finches and could describe the increase of the beak at the same time.

7.4.2 An Ecotoxicological SDM Example: Copper Changing the Size of Zooplankton

The conceptual diagram is shown in Figure 7.7. The model is presented by Jørgensen (2009) in Devellier (2009). The model software STELLA was used for the model simulation results. Copper is an algaecide causing an increase in the mortality of phytoplankton (Kallqvist and Meadows, 1978) and a decrease in the phosphorus uptake and photosynthesis. Copper also reduces the carbon assimilation of bacteria. The literature gives the change of the following three parameters in the model: growth rate of phytoplankton, mortality of phytoplankton, and mineralization rate of detritus with increased copper

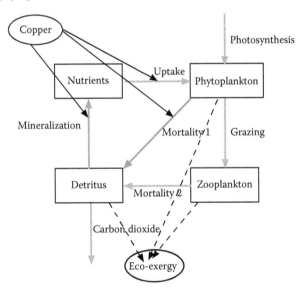

FIGURE 7.7 Conceptual diagram of an ecotoxicologial model focusing on the influence of copper on the photosynthetic rate, phytoplankton mortality rate, and mineralization rate. The boxes are the state variables, the thick grey arrows symbolize processes, and the thin black arrows indicate the influence of copper on the processes and the calculation of eco-exergy from the state variables. Due to the change in these three rates, it is an advantage for the zooplankton, and the entire ecosystem, to decrease its size. The model is therefore made structurally dynamic by allowing zooplankton to change its size, and thereby the specific grazing rate and the specific mortality rate according to the allometric principles. The size yielding the highest eco-exergy is currently found.

concentration (Havens, 1999). As a result, the zooplankton is reduced in size, which according to the allometric principles (for further details about these principles, see Chapter 8) means an increased specific grazing rate and specific mortality rate. It has been observed that the size of zooplankton in a closed system (a pond, for instance) is reduced to less than half the size at a copper concentration of 140 mg/m³ compared with a copper concentration of less than 10 mg/m³ (Havens, 1999). In accordance with the allometric principles (Peters, 1983), it would result in more than doubling the grazing rate and the mortality rate.

The model shown in Figure 7.7 was made structurally dynamic by varying the zooplankton size and using an allometric equation to determine the corresponding specific grazing rate and specific mortality rate. The equation expresses that the two specific rates are inversely proportional to the linear size (Peters, 1983). Different copper concentrations from 10 mg to 140 mg/m³ are found by the model in which zooplankton size yields the highest eco-exergy. In accordance with the presented SDM approach, it is expected that the size yielding the highest eco-exergy would be selected. The results of the model runs are shown in Figures 7.8–7.10. The specific grazing rate, the size yielding the highest eco-exergy, and the eco-exergy are plotted vs. the copper concentration in these three figures.

As expected, the eco-exergy even at the zooplankton size yielding the highest eco-exergy decreases with increased copper concentration due to the toxic effect on phytoplankton and bacteria.

The selected size (see Figure 7.9) is, at 140 mg/m³, as also indicated in the literature, less than half, namely, about 40% of the size at 10 mg/m³. The eco-exergy decreases from 198

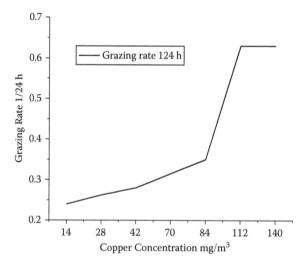

FIGURE 7.8 The grazing rate that yields the highest eco-exergy is shown at different copper concentrations. The grazing rate increases more and more rapidly as the copper concentration increases, but at a certain level, it is not possible to increase the eco-exergy further by changing the zooplankton parameters, because the amount of phytoplankton becomes the limiting factor for zooplankton growth.

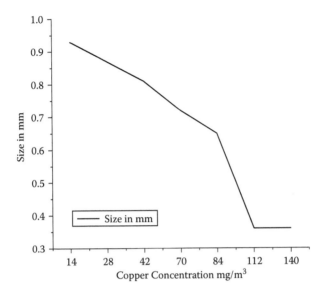

FIGURE 7.9 The zooplankton size that yields the highest eco-exergy is plotted vs. the copper concentration. The size decreases more and more rapidly as the copper concentration increases, but at a certain level, it is not possible to increase the eco-exergy further by changing the zooplankton size, because the amount of phytoplankton becomes the limiting factor for zooplankton growth.

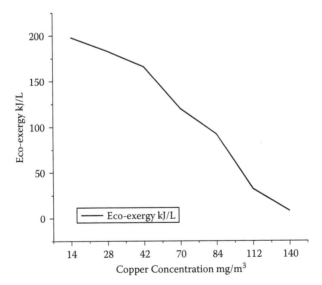

FIGURE 7.10 The highest eco-exergy obtained when varying the zooplankton size is plotted vs. the copper concentration. The eco-exergy is decreasing almost linearly with increasing copper concentration. The discrepancy from approximately a linear plot may be due to model uncertainty and discontinuous change of the copper concentration and the zooplankton size.

kJ/L at 10 mg/m^3 to 8 kJ/L at 140 mg/m^3. The toxic effect of the copper, in other words, results in an eco-exergy reduction to about 4% of the original eco-exergy level, which is a very significant toxic effect. If the zooplankton was not adaptable to the toxic effect by changing its size and thereby the parameters, the reduction in eco-exergy would have been even more pronounced at a lower copper concentration. It is therefore important for the model results that the model is made structurally dynamic, and thereby accounts for the change of parameters when the copper concentration is changed.

Zooplankton changes the size in the presented ecotoxicological case, and it is an advantage that SDMs can predict approximately the changes of the species' properties, but it is an even more important advantage that the state variables are predicted closer to the observations by the SDMs than by biogeochemical models because the organisms are able to adapt to the existing conditions. The toxic effect of copper would have been more pronounced if a nonstructurally dynamic model were applied, and that would inevitably have given too small concentrations of zooplankton.

> The 23 case studies that have been examined by SDM show that it is possible to capture and describe even quantitatively the observed structural changes—in the two illustrations presented here by the change of the beak size and the change of the zooplankton size. As ELT is a prerequisite for the application of SDM with eco-exergy as the goal function, the successful use of SDM can be considered a strong support for the ELT.

It is very important when we are using ecological models for development of prognoses to apply SDMs, because the model results are of course sensitive to the selection of parameters. The changes of parameters are giving, in most cases, completely different prognoses, and as has been shown in some of the case studies, SDMs are able to give prognoses that are closer to the real observation—have less uncertainty—than normal applied models. It is therefore concluded that SDMs are important to apply in ecological modeling that is able to account at least partially for adaptation and shifts in species composition.

7.5 The Compliance between ELT and Evolutionary Theories

Evolution is often described as a stepwise development from inorganic to simple organic molecules, and further on via more complex molecules and simple organisms toward more and more complex organisms that have more and more sophisticated properties. Even the simplest organisms would never have been formed spontaneously, because the probability that the right assembly of the complex biochemical compounds that determine the life processes is formed spontaneously would be so low that in practice it would never happen. Evolution has only been possible because the progressive steps were made one by one—from simple organic molecules to more and more complex molecules that were brought together to form something that, after numerous trials and errors, could be reproduced by maybe, in the firsthand, a very simple method. Later the reproduction method became more complex and more effective, as did all life-determining processes.

The evolution would, however, not have been possible without the genes, i.e., the ability to transfer information from generation to generation and thereby build on the shoulders of ancestors. The genes, via the complementary copies of mRNA molecules, determine the primary structure of the amino acid sequence of protein molecules—no more and no less. The primary structure of proteins determines all of the higher-order structures and thus their functions. Some proteins are used as building blocks, others play the role of enzymes, and another class of proteins participates in the so-called cell signaling, i.e., perceiving impulses (mostly chemical) from outside and elaborating them within the cell. More, there are also so-called motor proteins that transform chemical energy into mechanical energy. All these categories of proteins are indispensable for the development of the organism and its survival (Haugaard Nielsen, 2001).

The information about good solutions, which are solved by the genes that again put new constraints on survival, has to be transferred to the next generation: It is only possible to ensure survival in the light of the competition by use of genes. But the genes have also created new possibilities, because mutations and, later in the evolution, sexual recombinations and other mechanisms, which will be discussed later, create a wide range of new possible solutions. Therefore, as shown in Figure 7.11, what starts with

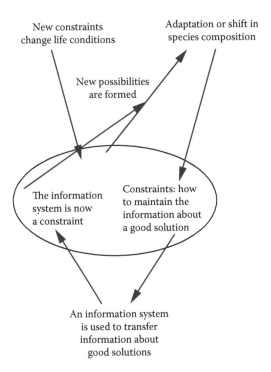

FIGURE 7.11 The life conditions are currently changed and have a high variability in time and space. This creates new challenges (problems) to survival. The organisms can adapt, or shifts to other, better-fitted species can take place. This requires an information system that is able to transfer the information about good solutions to the coming generations of organisms. Consequently, an information system is very beneficial, but it may also be considered a new constraint that opens up new possibilities.

constraints and new and better properties of the organisms or their ecological network end up as new possibilities through a well-functioning coding system.

The evolution can be considered an (almost) infinite shift between problems (constraints), followed by solutions that create new possibilities, but also new problems (constraints) that call for new solutions that create new possibilities, and so on—a staircase toward a more and more complex world. Notice, however, that the selection is taking place among phenotypes, while the new and better possibilities are created in the genes.

It has been widely discussed whether the evolution has been gradual or by jumps. The assumption today (see Jensen Peter, 2005) is that both descriptions are valid. During a period with stable conditions, a gradual evolution is dominant. When sudden changes in the conditions for life, for instance, sudden climatic conditions, occur, evolution by jumps may be more dominant.

The biochemistry of organisms is determined by the composition of a series of enzymes that again are determined by the genes. Successful organisms will be able to get more offspring than less successful organisms, and as the gene composition is inherited, the successful properties will be more and more represented generation after generation. This explains that the evolution has been toward more and more complex organisms that have new and emerging properties.

Figure 7.12, where an evolution index is plotted vs. time, illustrates this. The index is found as the product of the number of marine families and the β-value of the most developed organism at a given time. It is presumed generally that the development of the number of species follows the same pattern as the development of the number of families. The β-value has been presented in Jørgensen et al., 2005. The applied β-values are shown in Table 4.1. Table 7.2 gives the time of emergence applied in the figure.

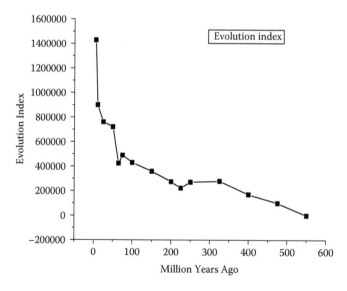

FIGURE 7.12 An evolutionary index is plotted vs. time. The calculation of the index is explained in the text.

TABLE 7.2 Time of Emergence Applied to
Calculate the Evolution Index Shown in Figure 7.12

Ma	Animals Emergent
500–550	A wide spectrum of invertebrates
475	Primitive fish
400	Fish
375	Amphibians
300	Reptiles
200	Mammals
30	Monkeys
10	Apes
1	Human

The evolution is formed by the constraints as challenges for the organism are originated in steadily changed life conditions. The organisms have been forced to provide most possible growth by a wide spectrum of changeable life conditions. Due to mutations, gene flow, and, later in the evolution, sexual recombinations, new solutions to survival in a changeable world have been provided, and if among the new solutions there are better solutions than the previous ones, the information has been stored in the genes and can be reused in the future. As the new solutions are developed from the previous available solutions, the number of families or species will increase. This image of evolution is in accordance with Monod's description of evolution, a result of chance and necessity.

The chance or random element of evolution is the steadily varying life conditions, and the necessity is the survival, because without survival there would be no continuation.

The genetic code contains the combination of four amino bases in blocks of four. The genetic code opens therefore for $4 \times 4 \times 4 \times 4$ combinations, but the code is used to select slightly over 20 amino acids, which implies that the code contains redundant signals. It does not matter at all for the efficiency of the code system, but it seems to indicate that the genetic code itself is a result of randomness.

New constraints are needed to give the evolution a kick from time to time. It has therefore probably been an advantage for evolution that the earth has been witness to several enormous natural catastrophes as, for instance, 65 million years ago, when an asteroid probably hit the earth and thereby created new conditions and therefore new challenges. New solutions may often not have a chance as long as old solutions are dominant. Only elimination of old solutions can give a new and different start. The evolution is dependent on catastrophes from time to time—volcanic eruptions, hurricanes, and sudden climatic changes. Compare with Holling's cycle (Holling, 1986); see Figure 6.4.

About 5 million years ago the climate in Africa changed dramatically. The precipitation was reduced significantly and the rain forest was replaced mainly in East Africa by savanna. This was a new challenge to the apes: it would be more beneficial for the apes in the savanna to be bipedal. That started the evolution toward modern man, *Homo sapiens*. The evolution toward a bigger brain volume was probably also or at least partially started randomly: Man became carnivorous and meat provided a wide spectrum of amino acids that supported brain growth. At the same time, a larger brain was required for man to form a coordinated team and become successful hunters. The tribe therefore formed an information network to facilitate the communication (exchange of information) among the members of the tribe and the hunting team.

The relationship between the random changes in the life conditions or the constraints and the evolution means that if we would repeat the evolutionary process, for instance, since the Cambrian explosion, it would inevitably follow the same ecological principles—the same ecosystem theoretical propositions, but it is not at all certain that the final results would be the same due to the role of randomness.

The entire evolution from the big bang via formation of the sun and the earth to the cultural-technological level of *Homo sapiens* of today has of course followed the second law of thermodynamics. This means that evolution, not surprisingly, is irreversible.

The irreversibility principle according to the second law of thermodynamics is a necessity for evolution, but not sufficient. Evolution has one direction: toward a higher and higher level of complexity, order, and information.

If the processes that make up evolution were reversible, evolution would not be directional.

The *biological* evolution from simple organic molecules to complex organic molecules to prokaryote cells to eukaryote cells to invertebrates to vertebrates and finally to *Homo sapiens* via mammals, monkeys, and apes shows the same direction of the evolutionary arrow: a higher and higher level of complexity and information.

Biological systems differ from physical systems by the increased complexity inherent in their development (Tiezzi, 2006).

Order and information can be measured by eco-exergy.

The mechanism proposed for evolution by ELT is that the eco-exergy steadily increases with some temporary decline of the eco-exergy levels when major catastrophes (completely new constraints) occur. The eco-exergy strives toward a higher and higher level under the prevailing conditions.

The forcing functions may change the conditions and therefore the eco-exergy level too. This explains, in Figure 7.12, the reduction in the evolution index.

Eco-exergy is increasing as a result of the evolutionary processes—more structure, more order, more information. Figure 7.13, reproduced from Jørgensen (2008), where more details are presented, shows a graph of the eco-exergy density for the most developed organisms as a function of time. In this context, it is important to remember that eco-exergy expresses biomass*information, and that it is the information factor that is increasing more and more rapidly. As previously discussed, the growth of biomass is limited in accordance with the amount of the limiting element, while the information still has the possibility to increase orders of magnitude. The evolution therefore has been

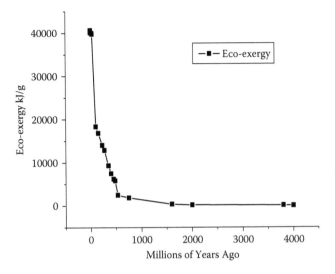

FIGURE 7.13 The eco-exergy density in kJ/g is plotted vs. the time. Notice the abrupt increase at the beginning of the Cambrian time, about 540 million years ago. Also, the enhanced increase by the emergence of mammals, hominoids, and humans can be seen clearly on the graph. See also Jørgensen (2008).

toward a more and more effective biochemical control, a more and more effective use of energy, and more and more sophisticated life forms, which is expressed thermodynamically in the figure. If semilogarithmic plots are made of the graph (Figure 7.14), the Cambrian explosion and the increase by the emergence of the first life are seen clearly. The eco-exergy density has increased close to exponentially, although the increase has been slightly faster than exponential, when life emerged and since the Cambrian explosion. This means that eco-exergy density grows in accordance with a first-order reaction:

$$(d \text{ eco-exergy density/vol}) = r \ dt \qquad (7.1)$$

where vol is the volume and r a rate constant. Figure 7.14 indicates that the slope is more than r at the emergence of the first primitive cell and after the Cambrian explosion. Particularly, the last few million years the slope is clearly more than the average slope, represented by the regression line.

The increasing diversity hypothesis, which is included in the evolutionary theories, states that the biosphere diversity is increasing through time. More species have been added to the environment as new ways of fitting into new ecological niches. Because organisms that can exploit all the possible resources better are favored or expressed differently, they are able to increase the eco-exergy by a better utilization of all the available resources, based on growth of the networks and the information. This is particularly emphasized in Figure 7.12, where the diversity and the size of the genomes are combined. There are several investigations showing the increase of the diversity over the last 600 million years to include the so-called Cambrian explosion (see, for instance, Figure 7.15). The number of plant species has increased enormously during the last 300 million years.

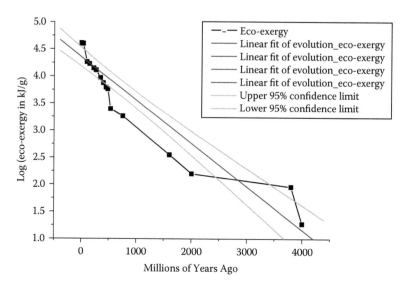

FIGURE 7.14 (See color insert.) Semilogarithmic plot of eco-exergy density vs. time. The plot is with a rough approximation linear (the red line), which implies that the eco-exergy density has increased exponentially. The correlation coefficient is 0.95. The confidential intervals ± one standard deviation are indicated by the green lines, The increase, however, has been particularly fast since the Cambrian explosion and just after the emergence of the first primitive life.

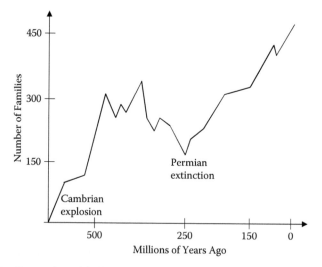

FIGURE 7.15 The number of shallow water marine families over the last 600 million years.

The number of vertebrate orders has in the same period increased from a couple to about 70. There are about 25 orders of mammals and more than 30 orders of birds.

The biodiversity has increased not only by the number of species, families, and orders, but also by the number of cell types, which also give more opportunities to exploit the resources more effectively. This is shown in Figure 7.16.

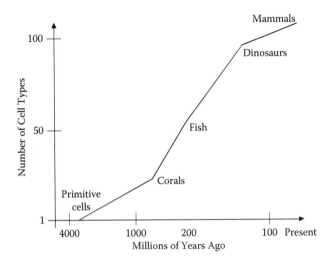

FIGURE 7.16 The graph shows the number of cell types that have evolved through geologic time. More details are presented in Jørgensen et al. (2005) and Jørgensen (2008).

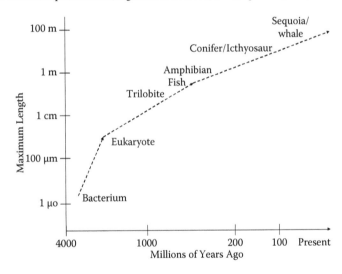

FIGURE 7.17 The maximum organism length has increased over the last 3,500 million years.

Another hypothesis is Cope's rule, which states that many lineages tend to increase in body size through time. The classic example is the evolution of the size of the horse. Besides hoofed animals, many dinosaur lineages and marine invertebrates show an evolution toward bigger size. Figure 7.17 shows the general increase of the maximum length of plants and animals during evolution. This is also in accordance with ELT. The bigger size means that less maintenance energy is used relative to the body size, because the respiration corresponding to the maintenance is proportional to the surface. If the length of an organism is increased by a factor of 2, the respiration is increased by a factor of 4, but the biomass by a factor of 8, which implies that the relative maintenance energy (kJ/kg body mass) is reduced to half. Example 7.3 finds the increase of the numeric eco-exergy efficiency for the evolution of the size of the horse.

Example 7.3

The horse was only 35 kg 45 million years ago. Today the horse is several hundred kg, and a horse could be 500 kg. There has obviously been an evolution toward a bigger size. The hypothesis is that the bigger size utilizes the food more effectively.

Find the eco-exergy efficiency of grass as food used for metabolism and growth as a function of the weight of horses. A model could be applied to answer the question.

Solution

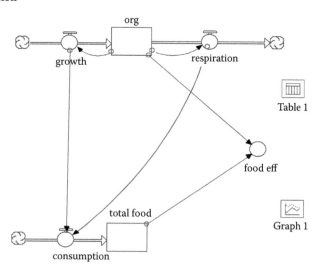

org is the horse that is growing and covering the maintenance of the complex structure far from thermodynamic equilibrium by respiration. The total food consumption covers the respiration and growth. A logistic growth should be applied to describe the growth, the growth should be proportional to the weight with the exponent 2/3, and respiration should be proportional to the weight with the exponent 3/4. The β-values applied for horse and grass are 2,127 and 200, respectively. The eco-exergy efficiency is found for the life span of the horse from 1 kg to the final weight. The following equations have been applied, using the STELLA format for the differential equations:

org(t) = org(t – dt) + (growth – respiration)*dt
INIT org = 1

Inflows:
Growth = 3*org^(0.67)*(1 – org/690)

Outflows:
Respiration = 0.5*org^(3/4)
Total_food(t) = total_food(t – dt) + (consumption)*dt
INIT total_food = 1

Inflows:
Consumption = growth + respiration
Food_eff = 2,127*100*org/(200*total_food)

The results of running the model to different maximum values are:

Final Weight	Efficiency of Food
500 kg	2.75%
400 kg	2.60%
300 kg	2.42%
200 kg	2.18%
100 kg	1.85%
50 kg	1.55%
35 kg	1.41%

As seen, the efficiency is clearly increasing by increasing body weight, which implies that it is beneficial for horses to increase their body weight.

7.6 Summary of Important Points in Chapter 7

1. The results of Chapters 6 and 7 can be summarized in an ecosystem core law (hypothesis) denoted the ecological law of thermodynamics (ELT) to indicate that thermodynamics is used to understand the dynamics of ecosystems:
   A system that receives a throughflow of exergy (high-quality energy, free energy) will try to utilize the exergy flow to move away from thermodynamic equilibrium, and if more combinations of components and processes are offered to utilize the exergy (free energy) flow, the system will select the organization that gives the system as much eco-exergy content (storage) as possible, i.e., maximizes dEx/dt.
2. The ELT is able to explain many ecological observations, and the following examples illustrate this use of ELT:
   a. Size of genomes
   b. Le Chatelier's principle
   c. The sequence of organic matter oxidation
   d. Formation of organic matter in the primeval atmosphere
   e. Photosynthesis by either C3, C4 pathways or CAM pathways

    f. Leaf size

    g. Biomass packing

    h. Cycling rates of nutrients

    i. Optimum foraging of the European starling

    More examples are given in Chapter 15.

3. By use of ELT it is possible to develop structurally dynamic models (SDMs) that can describe structural changes of ecosystems as a result of adaptation or changes in species composition. SDMs have worked successfully in 23 cases, as the models have been able to predict the observed structural changes with an acceptable uncertainty.

4. The *biological* evolution from simple organic molecules to complex organic molecules to prokaryote cells to eukaryote cells to invertebrates to vertebrates and finally to *Homo sapiens* via mammals, monkeys, and apes shows the same direction of the evolutionary arrow: higher and higher level of complexity, order, and information in complete accordance with the ELT.

5. The eco-exergy of the most developed organisms has increased at least exponentially during the time of evolution.

Exercises/Problems

1. Wallace hypothesized that insects that resemble in color the trunks on which they reside will be the best survivors. The peppered moth (*Biston betularia*) has two color forms, white moths peppered with black spots and a black form produced by a single allele. Between 1850 and 1920 the black form became dominant in England in response to the air pollution, because the dark form was better protected from predation by birds. Between 1950 and 1995 the white form again became dominant as a result of the air pollution abatement. Explain these observations by the use of the ELT.

2. Explain Figures 6.7 and 6.8 by the use of the ELT.

3. Is the prey-predator coevolution that is sometimes called the arms race consistent with the ELT?

4. Number of bird species vs. the latitude has with good approximation the following equation:

$$\text{Log number of bird species} = \text{latitude}/80 + (80 - \text{latitude})\ 2/70$$

A higher species diversity is presumed to give a better possibility to exploit all the ecological niches, and therefore the natural resources. Explain that the equation is consistent with ELT.

5. Species at a higher latitude have a tendency to have bigger body weight. Can this be explained by the ELT?

6. Prey abundance influences consumers' selectivity. Diets are broad when prey are scarce and narrow if food is abundant, as a function of time for searching for food. It is called the optimal foraging theory. Explain this theory by the ELT. (For further details about this theory, see Chapter 15.)

2

Properties
of Ecosystems

8

Ecosystems Are Open Systems

Openness creates gradients.
Gradients create possibilities.

All ecosystems are open to exchange of energy, matter, and information with their environment. The allometric principles and how to quantify openness are presented.

8.1 Why Must Ecosystems Be Open?

A 1 m tree that was planted more than 30 years ago may today be more than 30 m tall. It has increased the biomass many times and there may be more than a thousand times as many leaves today on the tree during the summer than 30 years ago. When we follow the development of an ecosystem over a longer period or during a couple of spring months, we are witness to one of the many wonders in nature: An inconceivably complex system is developing and growing in front of us. What makes this development of complex (and beautiful) systems in nature possible? In accordance with the classic thermodynamics, all isolated systems will move toward thermodynamic equilibrium. All the gradients and structures in the system will be eliminated and a homogenous dead system will be the final result, and the content of eco-exergy or free energy will be zero. As work capacity (eco-exergy) is a result of gradients in certain intensive variables such as the temperature, pressure, chemical potential, etc., a system at thermodynamic equilibrium will have no ability to do work. But trees and plants are moving in opposite directions. They are moving away from thermodynamic equilibrium at a faster and faster rate every year. We have already explained how this is possible. It is not a violation of the second law of thermodynamics, but the system is open and receives free energy from outside (mainly from solar radiation), and the free energy is used to maintain the system far from thermodynamic equilibrium. The openness is therefore an absolutely necessary condition for ecosystems. If they were not open, they would inevitably decompose to inorganic matter and would inexorably reach thermodynamic equilibrium.

With the concept of entropy Clausius reworded almost 150 years ago the second law of thermodynamics in a wider and more universal framework: *Die Entropie der Welt strebt einem Maximum zu* (The entropy of the world tends toward a maximum). Maximum entropy, which corresponds to the state of equilibrium of a system, is a state in which the work energy (free energy or exergy) is completely degraded and can no longer produce work. Free energy, exergy, and eco-exergy are all zero. Entropy is therefore a concept that shows us the direction of events. It has been called "time's arrow" by Harold Blum (Tiezzi, 2003). Barry Commoner (1971) notes that sand castles (order) do not appear spontaneously but can only disappear (disorder); a wooden hut in time becomes a pile of beams and boards: The inverse process does not occur. The spontaneous direction is thus from order to disorder and more entropy. This inexorable process has the maximum probability of occurring. In this way the concepts of disorder and probability are linked to the concept of entropy. Entropy is in fact a measure of disorder and probability even though, for ecosystems, it cannot be measured. Entropy generation can be calculated approximately, however, for reasonably complex systems (see Aoki, 1987, 1988, 1989).

To live and reproduce, plants and animals need a flow of energy. All vegetation, whether natural or cultivated, has been capturing solar energy for millennia, transforming it into food, fibers, materials, and work, and providing the basis for the life of the entire biosphere. The vast majority of the energy received by the earth's surface from the sun is dispersed: It is reflected, stored in the soil and water, used in the evaporation of water, and so forth. Only about 1% of the solar energy that falls on fertile land and waters is fixed by photosynthesis in plants (grass, trees, phytoplankton) in the form of high-energy organic molecules. By biochemical processes (respiration) the plants transform this energy into other organic compounds and work. Photosynthesis counteracts entropic degradation insofar as it orders disordered matter: The plant takes up disordered material (low-energy molecules of water and carbon dioxide in disorderly agitation) and puts it in order using solar energy. It organizes the material by building it into complex structures. Photosynthesis is therefore the process that, by capturing solar energy, decreases the entropy of the planet and increases its eco-exergy. The sun is an enormous engine that produces free energy and offers the earth the possibility of receiving large quantities of this free energy to be used for organization of life. Every year, the sun sends the earth $5.6*10^{24}$ joules of energy, or about 15,000 times more energy than we are consuming as fossil fuel every year.

Openness is a very important property for ecosystems, and it is therefore crucial to quantify this property. The next section will first present the allometric principles that are the basis for the quantification. Section 8.2 illustrates how it is possible to quantify the openness, which is important to explain the dynamics of ecosystems, particularly the dynamics of the various hierarchical levels of ecosystems.

8.2 The Allometric Principles and Quantification of Openness

Many process rates are, in physics, described as proportional to a gradient, a conductivity, or inverse resistance, and to the openness.

Compare for instance Fick's laws of diffusion and Ohm's law. The import and export from and to an ecosystem is therefore dependent on the differences between the ecosystem and the environment, as well as openness. For instance, the rate of the reaeration process of a water stream can be expressed by the following equation:

$$R_a = V \, dC/dt = K_a \, (T) \, A \, (C_s - C) \tag{8.1}$$

or

$$dC/dt = K_a \, (T) \, (C_s - C)/d \tag{8.2}$$

where R_a is the rate of reaeration, K_a is a temperature constant for a given stream, A is the area = V/d, V is the volume, d is the depth, C_s is the oxygen concentration at saturation, and C is the actual oxygen concentration. K_a is here the conductivity or inverse resistance. The faster the water flow in the stream, the higher is K_a. $(C_s - C)$ is the gradient and A is the area or the openness. Numerous expressions for rates in nature follow approximately the same linear equation.

The surface area of the species is a fundamental property for the species, because it indicates the contact surface between the organism and the environment. The surface area indicates quantitatively the size of the boundary to the environment, which can be utilized to set up useful relationships between size and rate coefficients in ecology. Loss of heat to the environment must, for instance, be proportional to the surface area and to the temperature difference, according to the law of heat transfer. The rate of digestion, the lungs, hunting ground, etc., are, on the one hand, determinants for a number of parameters (representing the properties of the species), and on the other hand, they are all dependent on the size of the organism. It is therefore not surprising that many rate parameters for plants and animals are highly related to the size, which implies that it is possible to get very good first estimates for most parameters based only upon the size. Naturally, the parameters are also dependent on several other characteristic features of the species, but their influence is often minor compared with the size, and good estimates based on the size can be used in many ecological models. It is possible, however, to take the shape of the organism into account by the use of a form factor = surface/volume, which may vary considerably among species.

The conclusion of these considerations is therefore that there should be many parameters (properties) that might be related to the size of the organisms.

There is, for instance, a strong positive correlation between size and generation time, Tg, ranging from bacteria to the biggest mammals and trees (Bonner, 1965); see Figure 8.1. This relationship can be explained by use of the relationship between size (surface) and total metabolic action per unit of body weight mentioned above. This implies that the smaller the organism, the greater the specific metabolic activity (specific indicates relative to the biomass or volume). The per capita rate of increase, r, defined by the exponential or logistic growth equations, respectively,

$$dN/dt = rN \tag{8.3}$$

$$dN/dt = rN(1^- - N/K) \tag{8.4}$$

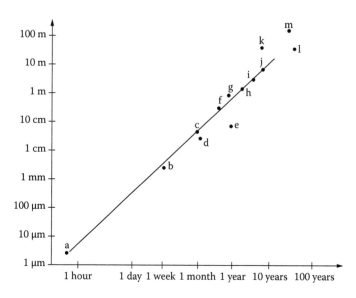

FIGURE 8.1 Length and generation time plotted on log-log scale: (a) *Pseudomonas*, (b) *Daphnia*, (c) bee, (d) house fly, (e) snail, (f) mouse, (g) rat, (h) fox, (i) elk, (j) rhino, (k) whale, (l) birch, and (m) fir. (From Peters, R.H., *The Ecological Implications of Body Size*, Cambridge University Press, Cambridge, 1983.)

is again inversely proportional to the generation time. This entails that r is related to the size of the organism, but as shown by Fenchel (1974), actually falls into three groups: unicellular, heterotherms, and homeotherms (see Figure 8.2).

The same allometric principles are expressed in the following equations, giving the respiration, food consumption, and ammonia excretion for fish when the weight, W, is known:

$$\text{Respiration} = \text{constant} * W^{0.80} \tag{8.5}$$

$$\text{Food consumption} = \text{constant} * W^{0.65} \tag{8.6}$$

$$\text{Ammonia excretion} = \text{constant} * W^{0.72} \tag{8.7}$$

It is expressed in the general equation (Odum, 1959):

$$m = k\,W^{-1/3} \tag{8.8}$$

where k is roughly a constant for all species, equal to about 5.6 kJ/g$^{2/3}$ day, and m is the metabolic rate per unit weight W.

Example 8.1

In an aquaculture basin with 2,000 fish with an average weight of 120 g the ammonium excretion is 800 g/24 h. It is important to keep the ammonium

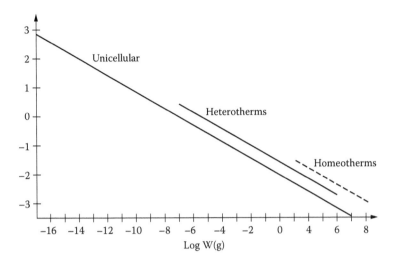

FIGURE 8.2 Intrinsic rate of natural increase against weight for various animals. (Data from Fenchel, T., *Oecologia*, 14, 317–326, 1974.)

concentration at 0.5 mg or below at the present pH, because ammonia but not ammonium is toxic to fish. Therefore, the water is exchanged at a sufficient rate to maintain 0.4 mg ammonium + ammonia/L. In another basin the aquaculture management wants to have 1,200 fish with an average weight of 1.4 kg. Which ammonium excretion should the management expect in this basin?

Solution

$800/2,000 = 0.4$ g/24 h $= k\ 120^{0.72}$
$k = 0.4/31.4 = 0.0127$ (1/24h)
Excretion by 1.4 kg fish $= 0.0127*1,400^{0.72}$
Excretion by 1.4 kg fish $= 2.34$ g/24 h
1,200 fish will therefore excrete 2,808 g/24 h.

Figures 8.3–8.5 give some ecotoxicological examples: the uptake rate of cadmium, the excretion rate for cadmium, and the concentration factor for cadmium (the concentration factor is the concentration in the organism divided by the concentration in water). The examples are just a few out of many in ecotoxicology. The shown relationships are, however, important, because it is possible according to the graphs to find an ecotoxicological parameter for any organism, provided that the parameters are known for two organisms and the size of the focal organism is known.

These considerations are all based on the allometric principles (Peters, 1983; Straskraba et al., 1997), which can be used to assess the relationship between the size of the units in the various hierarchical levels and the process rates, determining the need for the rate of energy supply. All levels in the entire hierarchy of an ecosystem are therefore due to the hierarchical organization, characterized by a rate that is ultimately constrained by their size.

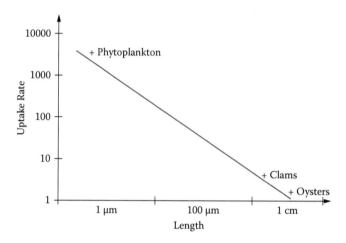

FIGURE 8.3 (See color insert.) Uptake rate (µg/g 24 h) plotted against the length of various animals (CD): (1) phytoplankton, (2) clams, and (3) oysters (From Jørgensen, S.E., *Ecol. Model.*, 22, 1–12, 1984.)

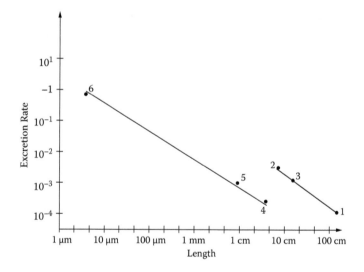

FIGURE 8.4 Excretion of Cd $(24\,h)^{-1}$ plotted vs. the length of various animals: (1) *Homo sapiens*, (2) mice, (3) dogs, (4) oysters, (5) clams, and (6) phytoplankton.

Example 8.2

Find the concentration factor for cadmium for a 2 m shark.

Solution

From Figure 8.5 it is possible (extrapolation) to read that the concentration factor will be approximately 80 for cadmium for a 2 m shark.

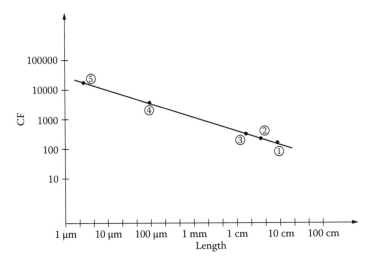

FIGURE 8.5 Concentration factor (CF) for Cd vs. size: (1) goldfish, (2) mussels, (3) shrimps, (4) zooplankton, (5) algae (brown-green).

TABLE 8.1 Relationship between Hierarchical Level, Openness (area/volume ratio), and Approximate Values of the Four Scale Hierarchical Properties, Presented by Simon (1973): Energy/Volume, Space Scale, Timescale, and Behavioral Frequency

Hierarchical Level	Openness[a,c] (A/V, m^{-1})	Energy[b] (kJ/m^3)	Space Scale[a] (m)	Timescale[a] (s)	Dynamics[c] (g/m^{3}s)
Molecules	10^9	10^9	10^{-9}	$<10^{-3}$	10^4–10^6
Cells	10^5	10^5	10^{-5}	10–10^3	1–10^2
Organs	10^2	10^2	10^{-2}	10^4–10^6	10^{-3}–0.1
Organisms	1	1	1	10^6–10^8	10^{-5}–10^{-3}
Populations	10^{-2}	10^{-2}	10^2	10^8–10^{10}	10^{-7}–10^{-5}
Ecosystems	10^{-4}	10^{-4}	10^4	10^{10}–10^{12}	10^{-9}–10^{-7}
Ecosphere	10^{-7}	10^{-7}	$2*10^7$	10^{13}–10^{14}	10^{-11}–10^{-12}

[a] Openness, spatial scale, and timescale are inverse to hierarchical scale.
[b] Energy and matter exchange at each level depend on openness, measured as available exchange area relative to volume. Electromagnetic energy as solar photons comes in small packages (quanta, hn, where h is Planck's constant and n is frequency), which makes utilization only at the molecular level possible. However, cross-scale interactive coupling makes energy usable at all hierarchical levels.
[c] Openness correlates with (and determines) the behavioral frequencies of hierarchical levels.

Openness is proportional to the area available for exchange of energy and matter, relative to the volume = the inverse space scale (L^{-1}). It may also be expressed as the supply rate = k × gradient × area relative to the rate of needs, which is proportional to the volume or mass. An ecosystem must, as previously mentioned, be open or at least nonisolated to be able to import the energy needed for its maintenance. Table 8.1 illustrates the relationship between the hierarchical levels, their openness, and the four scale hierarchical properties presented in Simon (1973). The openness is here expressed as the ratio of area to volume.

Many of the allometric characteristics are based on correlations between body size and other biological or ecological features of the organisms. These interrelationships are frequently comprehended as basic components of ecological hierarchies and basic objects of scaling procedures. Thus, they are highly correlated to hierarchy theory. Further details about this theory are presented in Chapter 9.

Following Simon (1973), hierarchy is a heuristic supposition to better understand complex systems, and following Nielsen and Müller (2000), hierarchical approaches are prerequisites for the definition of emergent properties in self-organized systems. Hierarchy theory is formulated by Allen and Starr (1982) and O'Neill et al. (1986). Kay (1984) has proposed an integrative concept of ecosystem-based classification and conception, which is compatible with most of the existing approaches to ecological system analysis. The theory has mainly been developed by Simon (1973), Allen and Starr (1982), and O'Neill et al. (1986), and recently there have been several applications in ecosystem analysis and landscape ecology. Further details are presented in Chapter 9.

Exchange of matter and information with the environment of open systems is in principle not absolutely necessary, thermodynamically, as energy input (nonisolation) is sufficient (the system is nonisolated) to ensure maintenance far from equilibrium. However, it gives the ecosystem additional advantages, for instance, by input of chemical compounds needed for certain biological processes or by immigration of species offering new possibilities for a better-ordered structure of the system. Therefore it can be concluded that

All ecosystems are open to exchange of energy, matter, and information with their environment.

All ecosystems exchange, for instance, water with the environment by precipitation and evaporation. The importance of the openness to matter and information is clearly illustrated in the general relationship between number of species, SD (species diversity), of ecosystems on islands and the area of the islands, A:

$$SD = C^*A^z \text{ (number)} \tag{8.9}$$

where C and z are constants. The perimeter relative to the area of an island determines how "open" the island is to immigration or dissipative emigration from or to other islands or the adjacent continent. The unit (L^{-1}) is the same as the above applied ratio of area to volume as a measure of openness.

Different species have very different types of energy use to maintain their biomass. For example, the blue whale uses most (97%) of the energy available for increasing the biomass for growth and only 3% for reproduction. Whales are what we call K-strategists, defined as species having a stable habitat with a very small ratio between generation time and the length of time the habitat remains favorable. This means that they will evolve toward maintaining their population at its equilibrium level, close to the carrying capacity. K-strategists are in contrast to r-strategists, which are strongly influenced by any environmental factor. Due to their high growth rate they can, however, utilize suddenly emergent favorable conditions and increase the population rapidly. Many fishes, insects, and other invertebrates are r-strategists. The adult female reproduces every season she is alive, and the proportion going into reproduction can be over 50%.

Example 8.3

The recovery time for ecosystems after major disturbances is related to the area disturbed by the following equation:

$$\text{Recovery times in years} = 10 \cdot (\text{area disturbed } (\text{km}^2))^{\wedge}(0.5)$$

A graph of the equation is shown below. For acidic rain and tsunamis, the recovery time is slightly less than shown in the equation, but for a lightning strike, autumn storm, land slide, oil spill, flood, volcanic eruption, and meteor strike the equation and graph give a very good approximation.

Explain why the equation and graph are in accordance with the allometric principles.

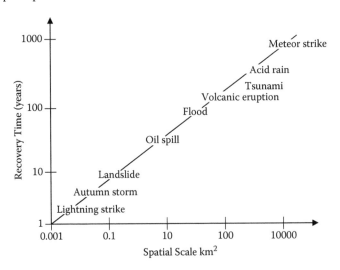

Solution

The recovery process should be proportional to the circumference of the disturbed ecosystems' area, because the recovery processes are coming from the environment of the ecosystems and the contact between the ecosystems and their environment is determined by the circumference. The circumference is proportional to the area in the exponent 0.5.

8.3 Summary of Important Points in Chapter 8

1. All ecosystems must be open to be able to get an energy flow from the environment, which is necessary to cover the free energy needed to maintain the ecosystem far from thermodynamic equilibrium. It is not sufficient for ecosystems to be open to be able to carry life. Eight conditions are needed for ecosystems to carry carbon-based life (see Section 5.6).
2. All ecosystems are open, and they can exchange energy, matter, and information with the environment.
3. The openness is defined as the circumference/area or the surface area/volume.
4. There is a clear relationship between openness (area/volume ratio) and approximate values of the energy/volume, space scale, timescale, and behavioral frequency (see also Table 8.1). The explanation is that many process rates are proportional to the contact possibilities between the ecosystem or the organism and the environment
5. Many properties of organisms are clearly related to the size of the organism. This is denoted in the allometric principles. For instance:

$$\text{Respiration} = \text{constant}^*W^{0.80}$$

$$\text{Food consumption} = \text{constant}^*W^{0.65}$$

$$\text{Ammonia excretion} = \text{constant}^*W^{0.72}$$

6. The allometric principles can be used to find a parameter (a property) for an organism by a log-log plot, which can be drawn by use of the process rate for two other organisms.

Exercises/Problems

1. Indicate the difference in development for an open ecosystem, a closed ecosystem (open to energy but only to energy), and an isolated ecosystem (completely closed to energy, matter, and information).
2. Calculate the ratio of the ammonium excretion rate for a 50 g and a 5,000 g fish.
3. A marine system with an area of 30,000 km^2 accidentally got a massive oil spill. How long will it take to recover? The recovery rate is dependent on the latitude, or rather the average temperature. Please indicate why.
4. What is the excretion and uptake rate of cadmium for a 5 cm mussel?
5. The concentration factor for PCB is 5,000 for a fish with the length 20 cm. What is the concentration factor for a seal that is 1.8 m long?
6. What is the generation time approximately for snails (10 cm)?
7. How much faster will a marine ecosystem with an area of 5 ha recover after a tsunami than a 50,000 ha marine system?

9

Ecosystems Have a Hierarchical Organization

Model every system, and if it is not modelable, make it modelable. (Sven Erik Jørgensen, changed from Galilei: Measure everything, and if it is not measurable, make it measurable.)

Ecosystems are organized hierarchically, which gives ecosystems several advantages: The variations (disturbances) are reduced on the higher and more important levels, repairs or adjustments of malfunctions are facilitated, the higher levels are less affected by environmental disturbances, and the ontic openness can be utilized. The openness determines the space and timescale for the hierarchical levels.

9.1 The Hierarchical Organization

The biological hierarchy is easy to observe. The biochemical processes take place in the cells, which have molecular components and structure to control the processes and protect the genome. In vertebrates, there are different types of cells, which are specialized to carry out the biochemical processes that are characteristic for different organs: the liver, the muscles, the kidneys, the heart, and so on. The cells that carry out the processes that take place in the liver make up the liver, and so on. It is a proper and effective hierarchical solution that the cells that have certain biochemical functions work together to ensure the functions of the organs.

The next hierarchical level after the organs is the individual organism. Organisms belong to different species according to their properties. Individuals of the species work together in populations that have numerous methods to ensure survival and growth for the individuals. The grazers form, for instance, a herd that makes it more difficult for the predators to attack the individuals of the herd. On the other side, the predators hunt together to obtain by cooperation a higher probability for successful hunting. Populations also use communication among the individuals to increase the probability for survival. Populations interact in a network and make up, together with the

155

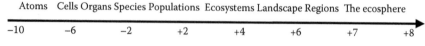

Approximate log (linear scale in meter) is indicated. Notice the axis is not linear.

FIGURE 9.1 The presented hierarchical levels are shown. Log to the linear scale in meters is indicated, but notice that the scale of the axis is not linear.

nonbiological component of the environment, the ecosystem. The interactions in networks have, as will be shown in Chapter 12, a synergistic effect that is able to increase the utilization efficiency of matter, energy, and information. Landscapes are formed by interactions among several ecosystems and regions comprised of many landscapes. The entire living matter on the earth makes up the biosphere, and the biosphere plus the nonbiological components are denoted the ecosphere.

The entire hierarchy is illustrated in Figure 9.1, and the figure indicates the corresponding spatial scale. Sometimes the organelle, a functional grouping of biomolecules, and the tissue, a functional grouping of cells, are included as levels between molecules and cells and between cells and organs, respectively.

The spatial scale of higher levels is naturally broader than the extent of lower levels. As the spatial scale determines the physical openness, it also determines the timescale and possible dynamics.

See also Table 8.1. The hierarchy is well fitted to the constraints that biological systems strive to move as far as possible away from thermodynamic equilibrium by the conditions of the thermodynamic laws, including ELT and the biochemical characteristics for life on earth.

Example 9.1

The openness of ecosystems in a landscape is determined to be $10^{-4.5}\,m^{-1}$. What is the openness (see also Tables 8.1 and 9.1) of landscapes and regions? Use Figure 9.1.

Solution

The openness is the inverse length, and the openness values of landscapes and regions are therefore, respectively, $10^{-6.0}$ and $10^{-7.0}\,m^{-1}$.

Higher levels change more slowly than lower levels due to the difference in spatial scale and the dynamics (see Table 8.1). It will be shown below, quantitatively, how the variations in one level of the hierarchy are averaged, and therefore result in fewer variations in the next level. The variation is, in other words, damped when we go up through the hierarchical levels, which is a clear advantage by the hierarchical organization. Variations and disturbances of the lower levels have a dynamic that may affect the lower levels in the hierarchy, but due to the hierarchical organization, the higher levels are much less affected because the variations caused by the disturbances are leveled out. Section 9.3 presents quantitatively the leveled out effect of the hierarchical organization, and Section 9.4 shows how the hierarchical organization and the dynamics on each level (see Table 8.1) are well fitted to the magnitude of the environmental disturbances as a function of the frequency. The magnitude of the environmental disturbances as a function of the frequency follows a power law. The recovery time (see the figure in Example 8.3) makes it possible for ecosystems to cope with the relationship between the frequency and the magnitude of the disturbances.

Another clear advantage of the hierarchical organization is that a malfunction of one level can easily be eliminated by replacing a few components in the lower level. The function of an organ can, for instance, be improved by a renewal of a few cells by the organisms, and an ecosystem may grow better by replacing one species with another one, that is better fitted to new emergent conditions.

9.2 Interactions between the Hierarchical Levels

Figure 9.2 shows how the processes on one level of the hierarchy determine the conditions in the next level, and how a level regulates and controls a lower level by feedback.

A specific level consists of interacting and cooperative entities that are integrated components in a higher organizational level.

The interaction of the entities in one level produces an integral activity of the whole, or expressed differently, the dynamics of a lower level generate the behavior of the higher level. The variation of a whole level is significantly smaller than the sums of the variation of the parts, but the degree of freedom of single processes is limited by the feedback regulation from the higher level.

Example 9.2

Give an example that illustrates that the population determines the framework of the individuals and that the individuals determine the properties and characteristics of the population.

Solution

A population of deer, for instance, determines in which direction the deer should move to get food and avoid predators. The individuals determine which type of food the population would prefer (could eat). The population sticks together to be better protected, which is particularly important for the young and weak deer. The individuals are carrying the genes that determine the properties and characteristics of the deer.

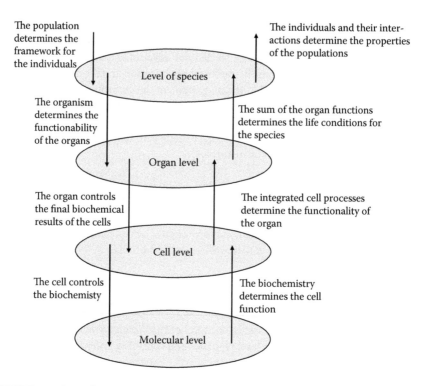

FIGURE 9.2   (See color insert.) The four lower levels of the ecological hierarchy are presented. The interactions between the levels, including the fifth level, are indicated.

It is often necessary by development of ecological models/models of ecosystems to consider several hierarchical levels. Identification of the level of organization and selection of the needed complexity of the model are not trivial problems. It is possible to indicate as many as 19 hierarchical levels in living systems, but to include all of them in an ecological model is of course an impossible task, mainly due to lack of data and a general understanding of nature. Only the nine levels shown in Figure 9.1 are, in most cases, considered in the development of ecological models. Usually, it is not difficult to select the focal level, where the problem is, or where the components of interest operate. The level one step lower than the focal level is often of relevance for a good description of the processes. For instance, the primary production of an ecosystem is determined by the processes taking place in the individual plants. The level one step higher than the focal level determines many of the constraints (see Figure 9.3). It is therefore often difficult to understand a particular behavior of an ecosystem at a particular level (see Allen and Starr, 1982) without also examining the behavior of the lower and the higher level.

Figure 9.4 illustrates a model with three hierarchical levels, which might be needed if a multigoal model is constructed. The first level could, for instance, be a hydrological model, the next level a eutrophication model, and the third level a model of phytoplankton growth, considering the intracellular nutrient concentrations. Each submodel will have its own conceptual diagram. The nutrients are taken up by phytoplankton in the

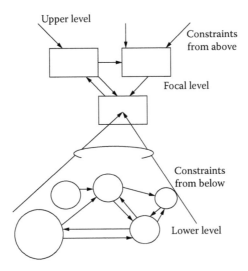

FIGURE 9.3 A focal level has constraints from both lower and upper levels. The lower level determines to a high extent the processes, and the upper level determines many of the constraints on the ecosystem.

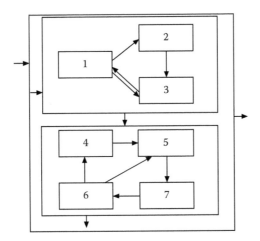

FIGURE 9.4 Conceptualization of a model with three levels of hierarchical organization.

third level model by a rate that is determined by the temperature, and nutrient concentration in the cells and in the water. The closer the nutrient concentrations in the cells are to the minimum concentrations in the cells, the faster is the uptake. The growth, on the other hand, is determined by solar radiation, temperature, and the concentration of nutrients in the cell. The closer the nutrient concentrations are to the maximum concentration, the faster is the growth. This description is according to phytoplankton physiology. Models that consider as well the distribution and effects of toxic substances might often require three hierarchical levels: one for the hydrodynamics or aerodynamics to account for the distribution, one for the chemical and biochemical processes of the toxic substances in the environment, and the third and last for the effect on the organism level.

Ecological models have been constructed for all the hierarchical levels from the cells to the ecosphere, except organs. It is in the ecological literature possible to find models of species, including their immediate environment, denoted habitats. There are also models between the two levels, populations and ecosystems, focusing on the interactions of either two or more species or the entire community. Ecosystem models often include several of the abiological factors influencing the populations of the ecosystems.

9.3 The Variations and the Hierarchical Organization

The variations and disturbances in one level of the hierarchy are damped or reduced in the next level of the hierarchy.

Statistics will be used to prove this statement. The standard deviation of an average of n components is the average deviation of the components divided by $n^{0.5}$. The cells contain 10^8 to 10^{10} molecules, and cells therefore have an environmental disturbance which is 10,000–100,000 times smaller than the disturbance of one molecule. The variation of the energy content, for instance, of molecules, may be 10% of the average energy content, or $0.1*10^9 = 10^8$ kJ/m^3, and the variation of the energy content of the cells is therefore 10,000 times smaller, or 10,000 kJ/m^3, which is 10% of the energy content of the cells. Similarly, for all the levels of the hierarchy, the number of components contained in the next level of the hierarchy ensures that the variations of intensive variables in each level are the same, expressed as percentage, provided that the focus is on intensive variables as the energy content and dynamics (see Table 8.1).

For extensive variables, it is different, which can best be illustrated by an example. If the ability of organisms to find food varies, for instance because of environmental disturbances, 100 kJ out of a total need of 1,000 kJ, or 10%, then the variation of a herd with 10,000 individuals to find food is 100 times less, or 0.1%, or 1, for an average organism, which means that for the entire herd it will be the average variation*the number of individuals in the herd = 1*10,000 kJ = 10,000 kJ. If the variation caused by the environmental disturbances would have been carried over to the next level of the hierarchy unchanged, the variation of the ability to find food for the herd would have been 100 kJ*10,000 = 1,000 MJ, or 100 times bigger.

Table 9.1 gives for the hierarchical levels presented in Table 8.1 an overview of the approximate damping effect of the variation due to disturbances. It is presumed that a variation on the molecular level is 100%, and the corresponding variations on each of the other six levels are indicated in the table as a percentage for an extensive variable.

It is found for the (n + 1)th level as the variation at the n-th level/(number of n-th level components contained in the (n + 1)th level)$^{0.5}$.

It is presumed that the number contained in the next hierarchical level is the ratio of the two scales in m^2, which of course should be considered a coarse approximation.

TABLE 9.1 Relationship between Hierarchical Level, Openness (Area/Volume Ratio), and Approximate Values of Variations or Disturbances, Provided the Variation on the Molecular Level Is 100%. Space scale (taken from Table 8.1). Number contained in the Next Level and the Value relative.

Hierarchical Level	Openness[a,c] (A/V, m^{-1})	Variation[b] (%)	Space Scale[a] (m)	Number in[c] Next Level	Value relative[d] (–)
Molecules	10^9	100	10^{-9}	10^8	100
Cells	10^5	0.01	10^5	10^6	10^4
Organs	10^2	10^{-5}	10^{-2}	10^4	10
Organisms	1	10^{-7}	1	10^4	10^{-4}
Populations	10^{-2}	10^{-9}	10^2	10^4	10^{-11}
Ecosystems	10^{-4}	10^{-11}	10^4	$4*10^6$	$4*10^{-18}$
Ecosphere	10^{-8}	$5*10^{-15}$	$2*10^7$	1	$2*10^{-34}$

[a] Openness, spatial scale, and timescale are inverse to hierarchical scale.
[b] Variation of the average contained component on the levels, assuming a 100% variation on the molecular level.
[c] Number of components contained in the next hierarchical level, assuming that the number is the ratio of the two space scales in the exponent 2.
[d] The value is on each level for an extensive variable, assuming that the value on the molecular level is 100.

This number is also indicated in the table. Moreover, the table shows the numerical *total* variation of the seven levels assuming that the molecular level has a variation of 100. It is indicated in the table as relative variation, because it is the numerical value of the total variation of *all* the components in a considered level relative to a value of 100 on the molecular level. It can be found from the following equation:

Total relative variation for level n + 1 = the variation of the average component of level n + 1 in %*the number of components*the total relative variation for level n/100.

It has been shown that more diverse communities have more stable ecosystem functioning and a consistently higher level of functioning over time than less diverse ones. More diverse plant communities have consistently higher productivity. Experiments over a period of seven years have given these clear results; see Allen et al. (2011). The results provide clear experimental support for the presented theoretical results in Chapter 9 of the damping effect from one hierarchical level to the next due to the number of components. The results support furthermore the advantages of a high biodiversity presented in Section 10.11.

Example 9.3

A landscape consists of 64 different ecosystems (see Figure 9.5). Find the variations of the average ecosystem, forming the landscape. What is the numerical total variation of the landscape, presuming that the variation on the molecular level is 100? Is the number of ecosystems in the landscape reasonable?

FIGURE 9.5   (See color insert.) Beautiful landscape in Rocky Mountains, Canada. The landscape consists of a lake, a river (not shown in the figure), wetlands in the littoral zone of the lake, a forest, a mountain ecosystem below the timber line, and a mountain ecosystem above the timber line. All the ecosystems influence each other and are open.

Solution

The variation on the ecosystem level is 10^{-11}%. The variation of the average of 64 ecosystems will be the square root of 64, or 8 times smaller or $1.25*10^{-12}$%. The total variation of the landscape is found as $64*1.25*10^{-12}$% $= 8*10^{-11}$%.

Landscapes have about 1/10 of the openness for ecosystems, and it should therefore be expected that landscapes very approximately contain about 100 ecosystems. Notice that the relative numeric total variation (disturbance, relative to the disturbance at the molecular level) increases from the molecular level to the cells, but decreases from the cells to the ecosphere. Table 9.1 shows clearly how the hierarchical organization damps variations and disturbances. The consequence is that ecosystems and particularly the ecosphere are very stable or, expressed differently, are very good survivors. Usually—at least before the massive impact by man—ecosystems survive hundreds of thousands of years, and the ecosphere has survived almost 4 billion years, as there has been life nonstop on the earth since the first primitive cells emerged 3.8 billion years ago.

9.4 The Frequency of Disturbances

The magnitude of variations, disturbances, and catastrophic events has a frequency according to the power law (see Bak, 1996).

The frequency, F, as a function of the magnitude, M, is therefore expressed by the following equation:

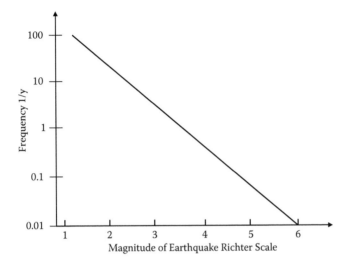

FIGURE 9.6 The frequency (events per year) is plotted vs. the earthquake magnitude for an earthquake-intensive area in the United States for a decade. Notice that the plot is double logarithmic, as the Richter scale expresses the magnitude of an earthquake in a logarithmic scale. As the graph is a straight line, the shown relationship must be a power function: frequency = constant times the earthquake magnitude in an exponent that can be shown to be –0.8.

$$F = a*M^b \tag{9.1}$$

An example is the distribution of earthquake magnitudes. Figure 9.6 shows the frequency vs. the earthquakes in Richter scale (which is logarithmic) in a district in the United States with many earthquakes during a decade. The graph is a straight line in a double logarithmic plot, which entails:

$$\text{Log } F = \log a + b \log M \tag{9.2}$$

where, as it can be seen from the graph (Figure 9.6), b = –0.8 and log a = 2.8 or a = 631.

The figure in Example 8.3 shows how the recovery time after a catastrophic event decreases with the area damaged. This means that the higher levels in the hierarchy ecosystems, landscapes, regions, and the entire ecosphere require a longer time to recover, but of course it is not surprising when a larger area is damaged and the openness determining the dynamic decreases with the area. Fortunately, the frequency of bigger catastrophic events decreases with the magnitude of the catastrophes. Figure 9.7 illustrates on the same graph the recovery time and the frequency, and it can be seen that they are opposite—they neutralize each other. The probability for a catastrophe that would damage large areas—ecosystems, landscapes, regions, and the entire ecosphere—therefore decreases with the hierarchical level. At the same time, as discussed in Section 9.3, the damping effect on the disturbance increases with the hierarchical level. The hierarchical organization is therefore beneficial for the stability (survival or maintenance)

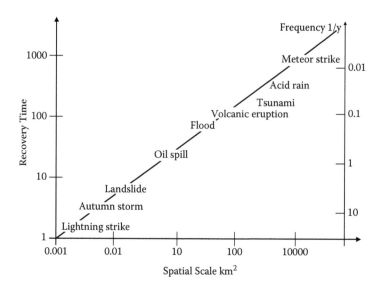

FIGURE 9.7 The approximate time for full recovery and the approximate frequency of cata-strophic events are indicated vs. the area damaged by the catastrophic event.

of the larger systems of the hierarchy. At the same time, the lower levels (see Table 8.1) have very high dynamics, which ensures a rapid renewal. The changes or the renewal is therefore a bottom-up effect, but due to the damping effect of the higher levels and the low frequency of disturbances, the survival of the higher levels, which is controlled by feedback from the lower levels, is ensured. The hierarchical organization is at the same time important for the renewal and development of nature and for the conservation of particularly the larger systems in nature.

Major catastrophes that have damaged significantly almost the entire ecosphere are known to have happened about 65 million, 200 million, and 251 million years ago. The frequency of a total or almost total destruction of the ecosphere is about one time per 100 million years. A very high percentage of all the species were extinguished by the three mentioned major catastrophes, and it took a very long time (according to fossils, more than 10,000 years before the ecosphere was at least partially recovered). This is in accordance with the trends in Figure 9.6. The last major meteor strike happened in Siberia in 1911, where more than 10,000 km² of forest was destroyed. The landscape is today not yet fully, but only partially, recovered. A meteor strike of this magnitude happens less than one time per century.

Example 9.4

In March 2011 there was an earthquake in Japan that measured 9.0 on the Richter scale. How often can such strong earthquakes be expected in Japan that have an area that is 50 times the intensive earthquake area valid for Figure 9.6? More than 1,000 km² of land were destroyed by that earthquake. How long would it take nature to recover after this earthquake?

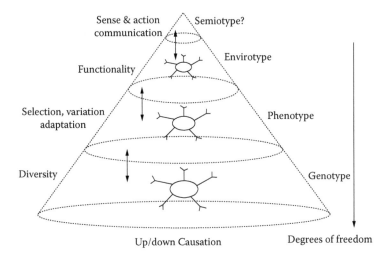

FIGURE 9.8 A biological hierarchy suggesting that interactions with the environment and finally the semiotics determine the development of the ontic openness of nature. The ontic openness increases from the genotype to the phenotype and further on to the envirotype and the semiotype, while the degrees of freedom decrease. (Reproduced from Jørgensen, S.E., et al., *A New Ecology*, Elsevier, Amsterdam, 2007. With permission.)

Solution

By application of extrapolation in Figure 9.6, it can be found that the frequency would be 0.0003. As Japan is also an earthquake-intensive area, it is presumed that the frequency is proportional to the area, which means that the frequency for Japan would be 0.0015, which is an earthquake about every 700 years. From Figure 9.6 we can read that the recovery would take about 400 years for mother nature.

9.5 Ontic Openness and the Hierarchy Theory

The cells include the genetic level apparatus. The number of nucleic acids or triplet codes that the genome combinations will exhibit is *immense*. This is also in accordance with the numbers applied by calculations of eco-exergy, based on the information content (see Chapters 4, 6, and 7 and Jørgensen et al., 2005). Ontic openness is definitely a reality at the cell level: There are enormous possibilities to form different genotypes. The number is beyond whatever can be determined, and therefore is in the hierarchical level ontic open. When we go to the populations (and partly to the ecosystems), there is an enormous number of possibilities to influence the genotypes, and therefore the phenotype level has an even larger ontic openness.

Patten (personal communication) has suggested another hierarchical level in addition to the genotype and phenotype levels, the exosomatic envirotype reflecting that an organism's genetic template and physiological manifestation are only realized with respect to its *ultimate* surroundings and the ecosystems, respectively. This is called the envirotype, and the ontic openness gets even larger. Recently, Nielsen (see Jørgensen et al., 2007) extended this view by adding a semiotic level above (Figure 9.8). This layer includes all kinds of communicative and cognitive processes, i.e., semiotics in the wide sense. This increases the number of possibilities further and represents the ultimate layer of realizing ontic openness, which increases when we go from ecosystems, to landscapes, to regions, and to the entire ecosphere. To summarize:

> The ontic openness increases in the hierarchy genotype-phenotype-envirotype-semiotype. It corresponds to an increasing ontic openness from the cells to the populations, to the ecosystems, and to the ecosphere.

Thus, at each layer of the biological hierarchy, we meet a new side of ontic openness. Interactions between hierarchical levels may, as indicated, take place in both upward and downward directions (compare also with Figure 9.2). The traditional view is that as we move up the hierarchy, we narrow the number of possibilities (the dynamic decreases, the physical openness decreases, and the damping effect increases). Therefore, as O'Neill et al. (1986) state, hierarchies are systems of constraints that are able to provide system regulations at steady-state conditions. Whenever rare events or system transformations occur the hierarchies are broken, and uncertainty takes place to a broad extent. Emergence due to ontic openness always exists, but it is just realized to a higher and higher extent when we move toward the higher hierarchical levels—in other words, what is not covered by the reductionistic view.

9.6 Summary of Important Points in Chapter 9

1. The ecological hierarchy is molecules, cells, organs, species, populations, ecosystems, landscapes, regions, and the ecosphere.

   A more practical model hierarchy would be molecules, cells, organs, species, habitats, populations, interactions between two or more populations, communities, ecosystems, landscapes, regions, and the ecosphere, or global models. It is possible in the literature to find models of all these levels.
2. The hierarchical organization has the following advantages:
   a. Cooperation of the component in a level.
   b. Possibilities to build a network by the component in a level to ensure a higher efficiency of the use of energy, matter, and information.
   c. A malfunction of one level can easily be reduced or eliminated by replacing a few components on the lower level. As the dynamic is higher at lower levels, this renewal can take place relatively rapidly.
   d. Disturbances (variations) are damped.
   e. The result of advantages a–d: The higher and therefore the more important a level is, the less vulnerable it becomes.
   f. A level is able to integrate the function of the lower level.
   g. The hierarchical organization makes it possible to utilize the ontic openness, which is decisive for the development.

    h.   A hierarchical distinction of different levels is the basis for recognizing emergent properties, which are the results of lower-level interactions and processes.

    i.   It can be helpful to overview and model complex systems.

3. The spatial scale of higher levels is broader than the extent of lower levels. As the spatial scale determines the physical openness, it also determines the timescale and possible dynamics.

4. The ontic openness increases in the hierarchy genotype-phenotype-envirotype-semiotype. It corresponds to an increasing ontic openness from the cells to the populations, to the ecosystems, and to the ecosphere.

Exercises/Problems

1. Which region is most vulnerable to pests?
   a.   A region consisting of only wheat fields totaling 5,000 km^2
   b.   A region consisting of 5 different landscapes and 88 different ecosystems
   Which of the two regions is most ontic open? The answers must be explained.

2. Indicate the hierarchy of models you would apply to model a region consisting of 5 landscapes and 68 ecosystems.

3. A landscape consists of 215 different ecosystems. Find the average variations of the ecosystem. It is assumed that the variations in the molecular level are 100. What is the numerical total variation of the landscape, presuming that the variation on the molecular level is 100? Is the number of ecosystems in the region reasonable?

4. How long will it take nature to recover from a catastrophe that has damaged completely 20,000 km^2 of land? How is mankind able to speed up the recovery process?

5. Explain why the hierarchical organization has been very important for evolution.

10

Ecosystems Have a High Diversity

Struggle for life is a permanent reality in nature and society.

Ecosystems select the best solutions to grow and develop as much as possible under the prevailing conditions, which vary enormously in time and space. Consequently, there is a need for many solutions to achieve this. It is shown how the wide spectrum of forcing functions, that vary in time and space, can be met by a wide spectrum of diversity in all the levels of the ecological hierarchy. It is discussed how this enormous diversity is important for the sum of buffer capacities (resistances) and for the spectrum of available buffer capacities. The high diversity in all levels of the ecological hierarchy can also explain that it is possible to find life in even the most extreme environment.

10.1 Introduction

The best solution to move as far away from thermodynamic equilibrium as possible under the prevailing conditions is continuously selected. The prevailing conditions, determined by the constraints, impacts, or forcing functions on the ecosystems, vary, however, enormously in time and space, as will be discussed in the next section. There is therefore continuous interaction between, on the one side, attempts to find the best solution for the most possible growth, and on the other side, the prevailing conditions, which vary in time and space. As a result and due to the possibilities by the use of the genes to remember good solutions, the selection leads inevitably to differentiation on all levels of the hierarchy.

The struggle between constraints or forcing functions and solutions has led to a high diversity, but as new solutions randomly are the result of mutations, sexual recombination, and transfer of genes between organisms, there is a probability that better and better solutions for growth will be selected. This implies that there is a high propensity to obtain more and more eco-exergy in accordance with ELT.

The process of finding better and better solutions by selection is denoted development, when the focus is on the short- or medium-term changes, while we would use the term *evolution* when we are focusing on the long-term results (see Section 7.5, where ELT was applied to explain evolutionary theories).

The next section uncovers the enormous variability of the constraints or forcing functions, which are the drivers of the diversity on all the hierarchical levels. From Section 10.3 to Section 10.10, the diversity in the hierarchical levels is presented: (1) in biochemistry—the molecular level, (2) in information carried by the cells— the genetic diversity, (3) of cells, (4) of organs, (5) of individuals in a population, (6) of species, (7) of communities and ecological networks, and (8) of ecosystems. Section 10.11 of the chapter looks into the advantages that diversity gives to the ecosystems, and Section 10.12 discusses why it is possible to find life in extreme environments on the earth.

10.2 The Wide Spectrum of Forcing Functions

All known life on earth resides in the thin layer enveloping the globe, known as the ecosphere. This region extends from sea level to about 10 km into the ocean, and approximately the same distance up into the atmosphere (Jørgensen et al., 2007). It is so thin that if an apple were enlarged to the size of the earth the ecosphere would be thinner than the peel. Yet a vast and complex biodiversity has arisen in the ecosphere, which can be considered an integrator of the particular elements favored by the biosphere (see Table 10.1). Table 10.1 illustrates the compositional differences between the spheres. Notice particularly that carbon is not readily abundant in the three abiotic spheres, yet is highly concentrated in the biosphere.

However, the conditions in the ecosphere for living organisms vary enormously in time and space.

The climatic conditions that are extremely important for life (see Chapter 5) vary enormously:

1. The temperature can vary from about –70 to about 55°C.
2. The wind speed can vary from 0 km/h to several hundred km/h.

TABLE 10.1   Atomic Compositions of the Spheres for the Seven Most Important Elements

	O	Si	Al	H	Na	Fe
Lithosphere	60.4	20.5	6.2	2.92	2.49	1.90
	N	O	Ar	C	Ne	He
Atmosphere	78.1	21.0	0.93	0.04	0.0018	0.0005
	H	O	Cl	Na	Mg	Ca
Hydropshere	66.4	33.0	0.33	0.28	0.034	0.006
	H	O	C	N	Ca	K
Biosphere	49.8	24.9	24.9	0.27	0.073	0.046

3. The humidity may vary from almost 0 to 100%.

4. The precipitation can vary from a few mm on average per year to several m/year, which may or may not be seasonally aligned.

5. Annual variation in day length is according to longitude from 0 to 24 h.

6. Unpredictable extreme events such as tornadoes, hurricanes, earthquakes, tsunamis, and volcanoes furthermore change the climate significantly.

The physical-chemical environmental conditions are:

1. Nutrient concentrations (C, P, N, S, Si, etc.)

2. Salt concentrations (important for both terrestrial and aquatic ecosystems)

3. Presence or absence of toxic compounds, whether they are natural or anthropogenic in origin

4. Rate of currents in aquatic ecosystems and hydraulic conductivity for soil

5. Space requirements and availabilities

The biological conditions are:

1. The concentrations of food for herbivores, omnivores, and carnivores

2. The density of organisms

3. The density of competitors for the resources (food, space, etc.)

4. The concentrations of pollinators, symbionts, and mutualists

5. The density of decomposers

The human impacts on natural ecosystems today add to this complexity.

The list of factors determining the life conditions is even longer, as we have only mentioned the most important forcing functions on ecosystems. In addition, the ecosystems have a history or path dependency, meaning that the initial conditions offer the possibilities of development. If we modestly assume that 100 factors define the life conditions, and each of these 100 factors could be on 100 different levels, then at least 10^{200} different life conditions are possible, which can be compared with the number of elementary particles in the universe, 10^{81}, or the number of seconds since the big bang, $5*10^{17}$. The confluence of path dependency and an astronomical number of combinations affirms that the ecosphere could not experience the entire range of possible states. Furthermore, its irreversibility ensures that it cannot track back to other possible configurations. In addition to these combinations, the formation of ecological networks (see Chapter 12) means that the number of indirect effects is magnitudes higher than the number of direct ones, and they are not at all negligible; on the contrary, they are often more significant than the direct ones, as discussed in Jørgensen et al. (2000) and Jørgensen et al. (2007).

What is the effect of this enormous variability of the constraints or forcing functions as a function of time and space? It is an enormous variability of the solutions to the problems caused by the forcing functions. It means that the diversity in all the hierarchical levels presented in Chapter 9 is expected to be very high.

The diversity in the levels of the hierarchy will be presented in Sections 10.3–10.10.

TABLE 10.2   Constituents in Living Matter

	Approximate Number of Different Compounds
Carbohydrates	>10,000
Nucleosides and nucleotides, constituents of nucleic acids	>10,000
Amino acids found in nature	48
Porphyrins	>10,000
Lipids	>10,000
Steroids	>1,000
Proteins Included the Enzymes on the Order of 10^{200} (or more)	
Hormones	>10,000
Metabolic intermediates	>1,000,000
Vitamins and vitaminoids	19

10.3 The Molecular Differentiation in Biochemistry

The differentiation of constituents and processes of living matter is enormous. Table 10.2 provides an overview of the most abundant constituents and the corresponding number of different molecules that can be found in the living matter. As we do not know how many species there are on the earth, it is of course even more difficult to give an exact number of compounds that can be found in all the living matter of nature. We are therefore forced to indicate an estimated number based on the number of compounds we have found and the estimated number of still unknown species. The number of amino acids in nature is indicated as 48, but only 28 can be found in the proteins of living matter. The remaining 20 are intermediate in biochemical processes. Twenty amino acids are directly encoded by the genetic code, and of these 20 amino acids, 8 are essential in human nutrition. Most amino acids found in living matter are L amino acids, but there are a few D amino acids in plants.

The number of different proteins in the biosphere is particularly high. Proteins have a molecular weight from about 10,000 to more than 1 million. If we presume an average molecular weight of 104,000 (Morowitz, 1968) and presume that the number of different proteins corresponds to the number of combinations of amino acids in the average protein to estimate the number of different proteins, we get about 10^{676} different proteins. An average protein contains 104,000/200 = about 520 amino acids, assuming an average molecular weight of amino acids of 200. This gives 20^{520}, or about 10^{676} different proteins. They are, of course, not all realized, but we have not included the eight amino acids that can be found in proteins but which are not encoded in the genes. We have also not included the few D amino acids that can be found in plants. We can, however, conclude that the number of possible proteins is astronomic. In the table, we have therefore indicated the very uncertain number: in the order 10^{200} corresponding to the number of combinations of constraints.

The nucleic acids determine the differentiation of DNA, which again determines the differentiation of proteins. The differentiation of the nucleic acids is due to a different sequence of amino bases. There are four amino bases, and three amino bases determine one amino acid in the protein chain. This gives 4*4*4 = 64 different possibilities of the amino base sequence to determine 20 amino acids. Therefore, several amino acids are determined by more than one amino base combination. The differentiation of nucleic acids is covered in the next section, where the focus is the genetic differentiation.

All in all, by this short overview of the biochemical differentiation it can be concluded that the number of different biochemical molecules is extremely high, and this implies that the number of possible biochemical processes among these components is even higher. The many combinations of constraints can therefore be met by the cells by a selection of biochemical components and processes among a very high number of possibilities.

10.4 The Genetic Differentiation

The genetic differentiation is the differentiation of the information that controls the biochemical processes. The information carried by the cells and contributing to the eco-exergy of the organisms (see Chapters 6 and 7 and Jørgensen et al., 2007) is stored in the genomes. Eco-exergy is exergy or work capacity found in ecological context for the components of an ecosystem, with the same ecosystem at thermodynamic equilibrium as the reference state. The information is in the form of the amino base sequence of the nucleic acids in DNA and RNA. Table 10.3 provides an overview of the results of the

TABLE 10.3 Genome Size and β-Values Found on the Ongoing Whole-Genome Sequencing Project

Organisms	Genome Size (Mb)	Repeat %	β-Values
Human	2,900	46	2,173
House mouse	2,500	38	2,127
Tiger fish	400	9	499
Sea squirrel	155	10	191
Malaria mosquito	280	16	322
Fruit fly	137	2	184
Nematode worm	97	0.5	133
Human malaria parasite	23	0.5	31
Rodent malaria parasite	25	0.5	34
Amoeba	34	0.5	46
Intracellular parasite	34	0.5	46
Brewery yeast	12	2.4	16
Fission yeast	14	0.35	19
Microsporidium parasite	2.5	<0.1	3.4
Mustard weed	125	14	147
Rice	400	50	275
Virus	0.01	0	1.01
Viroid	0.0036	0	1.0004

whole-genome sequencing project. The genome's size is indicated in Mb and the repeat of the sequence in percentage. The β-values that cover the eco-exergy for various organisms in the unit detritus equivalent per g biomass are also included in the table:

$$\text{Ex-total} = \sum_{i=1}^{N} \beta_i c_i \text{ (as detritus equivalent)}$$

The β-values in Table 10.3 are found from the knowledge of the entire genome for the 16 organisms (see Jørgensen et al., 2000, 2007). Table 4.1 gives a more complete list of β-values.

As quantification of the genomic differentiation for the various organisms, we can apply either the size of the genome, the number of the nucleotides or amino acids encoded by the genomic code, or the β-values or eco-exergy/g biomass. The genomic differentiation for humans corresponds to $1.590 * 10^9$ nucleotides, which can encode for about 500 million amino acids, as three nucleotides are needed to encode 1 amino acid.

Example 10.1

Approximately how many amino acids are in 100 mg of proteins (enzymes)? How much more information is contained in the amino acid sequence of the amino acids in 100 mg of proteins than in the human nucleotides? What are the possibilities to increase the information of living organisms by a continuous evolution?

Solution

Let us presume an average molecular weight of amino acids of 200 g, which means that 100 mg will contain 1/2,000 mole of amino acids, or A (Avogadro's number)/2,000 = $6.2 * 10^{23}/2,000 = 3.1 * 10^{20}$ amino acids, or about $6 * 10^{12}$ times more than the nucleotides in the genomes. This means that there are almost no limits to how much the information of organisms can increase.

10.5 The Diversity of Cells

The number of distinct cell types in the adult human body includes several hundred, almost a thousand, different and distinct types. The number of cell types has increased during evolution, and the most advanced organisms—the mammals—have the most cell types; see also Jørgensen et al. (2005).

Table 5.2 gives an overview of the differences between the two main types of cells: prokaryote and eukaryote cells that represent, respectively, the most simple cell type that emerged about 3,500 million years ago and a very advanced cell type that emerged about 2,000 million years ago. The difference between the two types also represents very well the spectrum of cell properties. The COPE homepage gives a detailed overview of the various cells types and their properties.

10.6 The Diversity of Organs

An organ is a collection of tissues joined in a structural unit to serve a common function. Usually there is a main tissue and sporadic tissues. The main tissue is the one that is unique for the specific organ. For example, the main tissue in the heart is the myocardium. Functionally related organs often cooperate to form whole organ systems. Organs exist in all higher biological organisms, and they are not restricted to animals, but have also been identified in plants.

Mammals have the highest number of different organs. There are 11 major organ systems found in mammals, encompassing about 200 organs:

Circulatory system: Pumping and channeling blood to and from the body and lungs with heart, blood, and blood vessels.

Digestive system: Digestion and processing food with salivary glands, esophagus, stomach, liver, gallbladder, pancreas, intestines, rectum, and anus.

Endocrine system: Communication within the body using hormones made by endocrine glands, such as the hypothalamus, pituitary or pituitary gland, pineal body or pineal gland, thyroid, parathyroids, and adrenals, i.e., adrenal glands.

Excretory system: Kidneys, ureters, bladder, and urethra involved in fluid balance, electrolyte balance, and excretion of urine.

Integumentary system: Skin, hair, and nails.

Lymphatic system: Structures involved in the transfer of lymph between tissues and the bloodstream, the lymph and the nodes, and vessels that transport it, including the immune system: defending against disease-causing agents with leukocytes, tonsils, adenoids, thymus, and spleen.

Muscular system: Movement with muscles.

Nervous system: Collecting, transferring, and processing information with brain, spinal cord, peripheral nerves, and nerves.

Reproductive system: The sex organs, such as ovaries, fallopian tubes, uterus, vagina, mammary glands, testes, vas deferens, seminal vesicles, prostate, and penis.

Respiratory system: The organs used for breathing, the pharynx, larynx, trachea, bronchi, lungs, and diaphragm.

Skeletal system: Structural support and protection with bones, cartilage, ligaments, and tendons.

10.7 Diversity among Individuals

It is an impossible task to indicate the differentiation of individuals in all the species on the earth, simply because we do not know all the species (about an estimation of the number of species, see Section 10.8). We can, however, indicate the differentiation in the human population and thereby get an indication of the individual differentiation. *Homo sapiens* has $1.590*10^9$ nucleotides, and 99% of these we share with other members of the human population. If we multiply the number of nucleotides that we do not share with other (1%) = $1.590*10^7$ nucleotides with 6.5 billion, which is the human population (2011), we do not get the number of different nucleotides, because some of the nucleotides are shared not with everybody but with some (few) individuals. The multiplication $1.590*10^7$ nucleotides times $6.5*10^9 = 1.033*10^{17}$ gives, however, an indication of the diversity that can be found in the human population.

It can be concluded that the diversity among individuals of the same species is huge. If we account for the diversity of all individuals in an ecosystem, the diversity becomes astronomical, and of course, if we consider all species (estimated to be 5 million), the diversity on the level of individuals increases correspondingly.

10.8 Species Diversity

A species is a basic unit of biological classification and a taxonomic rank. A species is often defined as a group of organisms capable of interbreeding and producing fertile offspring. While in many cases this definition is adequate, more precise or differing measures are often used, such as similarity of DNA, morphology, or ecological niche, which is defined as the ecological conditions under which species can live and reproduce. In the context of differentiation, the similarity of DNA explains why individuals of the same species can survive the same combination of constraints. We know about 2 million species, but because we still, in untouched nature, particularly the tropical rain forest, find new species, it is estimated that the number of species is much higher than 2 million. As already mentioned, most estimations point toward about 5 million species.

The differentiation of species is expressed by several species diversity measures:

1. Number of species S eventually expressed as log S.
2. Shannon's index of general diversity: $H = -\Sigma n_i/N \log n_i/N$, where n_i is the number of individuals of the i-th species and N is the total number of individuals. The natural logarithm can also be applied.
3. Evenness index $e = H/\log S$.

Species diversity is used to express the possibilities according to the above-presented discussion to maintain the functioning of ecosystems by changed forcing functions (constraint). The idea is that the species diversity represents the diversity of characters or properties of the organisms to meet the changed conditions (constraints). Many ecological investigations have shown that (see Gaston, 2000) the two diversities are related. Figure 10.1 illustrates the result of an examination of 218 bird species (Gibbons et al., 1993). The richness of characters or properties is less than the species richness, but there is, as shown, a close relationship between the diversity of properties and the diversity of species.

Example 10.2

An ecosystem is dominated by 5 species and has in total 405 species. The total number of individuals in the ecosystem (microorganisms are not accounted for in this context) is 100,000. The numbers of individuals in the five dominant species are 8,000, 20,000, 20,400, 22,600, and 25,000. The 400 not dominating species have with good approximation 10 individuals each. Find the values of the three diversity indices proposed in the text.

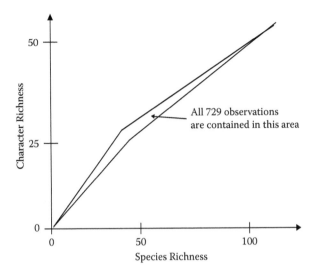

FIGURE 10.1 The result of an examination of 218 bird species is illustrated (Gibbons et al., 1993). The richness of characters or properties is less than the species richness, but there is, as shown, a close relationship between the diversity of properties and the diversity of species.

Solution

The number of species is S = 405 or log S = log 405 = 2.603.

H = –(0.08*log 0.08 + 0.2*log 0.2 + 0.204*log 0.204 + 0.226*log 0.226
+ 0.25*log 0.25 + 400*0.0001*log 0.0001) = –(0.08*(–1.1) + 0.2*(–0.7) +
0.204*(–0.690) + 0.226*(–0.656)+0.25*(–0.6) + 400*0.0001*(–4)) = 0.654

E = 0.654/2.603 = 0.251

Both the diversity and the evenness are small. The diversity for many ecosystems is between 2 and 4.

The concept of ecological niches describes for various species the ranges of forcing functions (conditions) that are required for the considered species to survive. It can be turned around: an organism or population responds to the distribution of resources and competitors and how it is able to alter these factors.

The competitive exclusion principle (Gause, 1934) claims that when two or more species are competing over the same limited resource, only the best one will survive. The contrast between this principle and the high number of species has, for a long time, been a paradox. The explanation is rooted in the enormous variability in time and space of the conditions and in the variability of a wide spectrum of species' properties. A competition model, where three or more resources are limiting, gives a result that is very different from cases where one or two resources are limiting. Due to significant fluctuations in the different resources, it can be shown by use of a competition model that many species competing about the same (wide) spectrum of resources can coexist. It is therefore not

FIGURE 10.2   (See color insert.) One square meter of a natural ecosystem on an island in the Adriatic Sea. It was possible during a period of 2 h to count 92 plant and insect species.

FIGURE 10.3   (See color insert.) The photo shows a shallow lake or wetland 13 km north of Copenhagen. The freshwater ecosystems in the temperate zone often have a very high diversity, and they offer many ecosystem services to society (see also Table 10.4).

surprising that there exist many species in an environment characterized by an enormous variation of both abiotic and biotic factors. Figures 10.2 and 10.3 illustrate the high diversity that we find in natural ecosystems. It is in accordance with the niche concept and the exclusion principle.

10.9 Differentiation of Communities and Ecological Networks

The species are operating in ecosystems with other species and are integrated in ecological networks. The network has synergistic effects (see Patten, 1992; Jørgensen et al., 2007). There are many possible networks available to select from by the ecosystem to obtain the highest possible synergistic effect of the network and the highest possible eco-exergy of the ecosystem. Most ecosystems are inhabited by at least 5,000 species of plants, animals, and bacteria. An ecological model of an ecosystem includes rarely more than 50 (dominant) species as state variables, which are interacting in an ecological network covering in most cases more than 90% of the cycling of matter, energy, and information. Most species contribute not very much to the cycling of matter, energy, and information and are just waiting in the wings to take over in case the prevailing conditions become favorable for them. But 5,000 species are able to form 5,000*4,999*4,998, ..., 4,951 different networks, or about 10^{184} possible networks. These calculations have a high uncertainty, because the number of species in different ecosystems varies and the number of species that should be included to cover >90% of the recycling of matter, energy, and information will also differ considerably from case to case, but the main conclusion is not the number, but that the number of possible networks in ecosystems is enormous. Furthermore, the networks vary in time and space, because the species are changing their properties, and the interactions among species, as expressed in the network, will therefore also inevitably vary significantly in time and space. See, for instance, the description of the arms race between prey and predators (see Covitch, 2010).

10.10 Diversity of Ecosystems

Different prevailing, mainly climatic conditions are, together with the dominant, mainly plant species, the basis for a classification of ecosystems. Table 10.4 gives an overview of the different ecosystems, their annual production of biomass, their annual production of work energy (eco-exergy), and the most important criteria for the classification. The differentiations among the ecosystems are very significant, in conditions, biomass production, and eco-exergy production.

The classification of ecosystems is usually based on our visual perception of nature, and the table can easily be expanded. For instance, wetlands encompass bogs, ferns, forested wetlands, shallow lakes, wet meadows, marshes, and reed swamps. You can distinguish between rivers and streams, and lowland river and streams, and mountain rivers and streams—just to mention some possibilities to enlarge the list of ecosystems. The number of biomes, i.e., natural systems offering different life conditions, perhaps covers better the variability of ecosystems. Globally, the number of different biomes is about 400–500 (Gaston, 2000).

Ecosystems increase their differentiation when they develop from an early stage to a mature stage; see Odum's attributes (Odum, 1968) and the three growth forms presented in Jørgensen et al. (2000, 2007).

Ecosystems encompass an enormous diversity in the form of many species, many possible networks, an astronomic number of amino base sequences in the DNA,

TABLE 10.4   Ecosystems, Their Typical Annual Biomass, Eco-Exergy Production, and Characteristics

	MJ/m² year as Biomass	GJ/ha year Eco-Exergy	Characteristics
Desert	0.9	2,070	Aridity, water is the limiting factor
Open sea	3.5	2,380	Sea with minor influence from land
Coastal zones	7.0	4,830	Sea with major influence from land
Coral reefs, estuaries	82	1.0 million	Calcium carbonate skeleton engineered by coral
Estuaries	75	950,000	Interface of marine, freshwater, and terrestrial systems
Lakes, rivers (freshwater)	11	93,500	Freshwater systems surrounded by terrestrial systems
Coniferous forests	15.4	539,000	High density of coniferous trees
Deciduous forests	26.4	1.0 million	High density of deciduous trees
Temperate rain forests	39.6	1.5 million	Forest in temperate climate with high rainfall
Tropical rain forests	82	3.0 million	Forest with high temperature, rainfall, and diversity
Tundra	2.6	7,280	Low temperature, strong winds, low precipitation
Croplands	20.0	420,000	System managed for food production
Grassland	7.2	18,000	Grass species dominant vegetation
Wetlands	18.0	45,000	Intermediate between terrestrial and aquatic ecosystems

and the amino acids in proteins. This enormous diversity makes it possible for ecosystems to find a solution to survive, grow, and develop under the pressure of highly variable constraints.

Let us, however, give a description of some of the life forms in some of the most extreme environments, because that would give us a better understanding of how constraints provoke differentiation.

10.11 The Advantages of a High Biodiversity

Biodiversity is defined as the diversity of all living beings. It has three levels (Leveque and Mounolou, 2003):

1. The genetic variability of populations, denoted intraspecific diversity, which explains the response of species to changing forcing functions.
2. Diversity among species in terms of their ecological functions. This is the prerequisite for selection and adaptation, which is driving the evolution.
3. Ecosystem diversity, which is the variability in space and time of the habitats. This is the driver of the global cycling of biologically important elements.

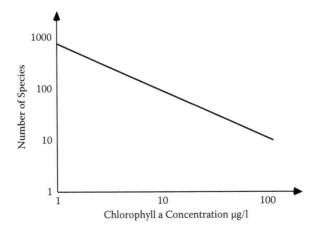

FIGURE 10.4 Weiderholm (1980) obtained the relationship shown for a number of Swedish lakes between the number of species and eutrophication, expressed as chlorophyll-a in µg/L.

O'Neill et al. (1986) examined the role on heterotrophs of resistance and resilience and found that only small changes in heterotroph biomass could reestablish system equilibrium and counteract perturbations. He suggests that the many regulation mechanisms and spatial heterogeneity should be accounted for when the stability concepts are applied to explain ecosystem responses.

These observations explain why it has been very difficult to find a clear relationship between ecosystem stability in its broadest sense and species diversity. Compare also with Rosenzweig (1992, 1996) and May (1973), where almost the same conclusions are drawn.

It is observed that increased phosphorus loading gives decreased diversity, (Ahl and Weiderholm, 1977; Weiderholm, 1980), but very eutrophic lakes are very stable. Figure 10.4 gives the result of a statistical analysis from a thousand Swedish lakes. The relationship shows a correlation between number of species and the eutrophication, measured as chlorophyll-a in µg/L. A similar relationship is obtained between the diversity of the benthic fauna and the phosphorus concentration relative to the depth of the lakes.

Model studies (Jørgensen and Mejer, 1977, 1979; Jørgensen, 2002) have also revealed that in lakes with a high eutrophication level, minor changes to nutrient inputs are obtained by a relatively small diversity. The low diversity in eutrophic lakes is consistent with the above-mentioned results by Ahl and Weiderholm (1977) and Weiderholm (1980). High nutrient concentrations = large phytoplankton species. The specific surface does not need to be large, because there are plenty of nutrients. The selection or competition is not (any longer) on the uptake of nutrients, but rather on escaping the grazing by zooplankton, and here greater size is an advantage. The spectrum of selection becomes more narrow, which means reduced diversity.

If a toxic substance is discharged to an ecosystem, the diversity will be reduced. The species most susceptible to the toxic substance will be extinguished, while other species, the survivors, will metabolize, transform, isolate, excrete, etc., the toxic substance, and thereby decrease its concentration. We observe a reduced diversity, but simultaneously will further input of toxic substances only cause relatively small changes, because the species present are able to cope with the impacts from the toxic substances? Model studies of toxic substance discharge to a lake (Jørgensen and Mejer, 1977, 1979) demonstrate the same inverse relationship between the buffer capacity to the considered toxic substance and diversity.

Another consequence of the complexity of ecosystems mentioned above should be considered here. For mathematical ease, the emphasis has been, particularly in population dynamics, on equilibrium models. The dynamic equilibrium conditions (steady state, not thermodynamic equilibrium) may be used as an attractor (in the mathematical sense, the ultimate ecological attractor is the thermodynamic equilibrium) for the system, but the equilibrium will never be attained. Before the equilibrium should have been reached, the conditions, determined by the external factors and all ecosystem components, have changed and a new dynamic equilibrium, and thereby a new attractor, is effective. Before this attractor point has been reached, new conditions will again emerge, and so on. A model based upon the equilibrium state will therefore give a wrong picture of ecosystem reactions. The reactions are determined by the present values of the state variables, and they are different from those in the equilibrium state. We know from many modeling exercises that the model is sensitive to the initial values of the state variables. These initial values are a part of the conditions for further reactions and development. Consequently, the steady-state models may give other results than the dynamic models, and it is therefore recommended to be very careful when drawing conclusions on the basis of equilibrium models. We must accept the complication that ecosystems are dynamic systems and will never attain the steady-state equilibrium, because the conditions are changed before the steady-state equilibrium is reached. We therefore almost always need to apply dynamic models as widely as possible, and it can easily be shown that dynamic models give other results than static ones.

It can be concluded that there are very complex relationships between the diversity, stability, and resistance on the one side and the impacts of forcing functions on the other side. Diversity is, however, important for the ability of ecosystems to meet new and (unexpected) forcing functions with small changes. If a toxic substance, for instance, is discharged to an ecosystem, there is a higher probability that the primary production can be maintained in spite of the toxic substance if there are more plant species present in the ecosystem. There is a higher probability of finding species that can resist the toxic substance if there are more plant species to select from. The results of the Tilman and Downing (1994) grass experiments support the presented hypothesis, that more species give a higher probability to cope with a forcing function. They have shown that drought resistance increases with the number of species. Each additional plant, however, contributes less (see Figure 10.5). It can therefore be concluded that there is not a simple relationship between diversity and stability or even resistance, but a higher species diversity means an increased probability for the ecosystem to cope with a wider spectrum of forcing functions, or expressed differently: A higher species diversity implies that there is a higher probability that the ecosystem can meet a (new and unexpected) forcing function by less radical changes. It is also consistent with the results obtained by Zavaleta et al. (2010) by an empirical model—an empirical: higher species richness is required to provide multiple ecosystem functions.

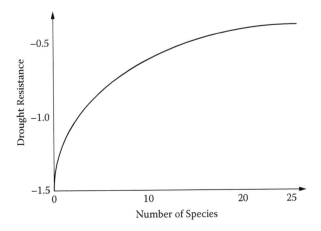

FIGURE 10.5 The results of the experiments in grassland by Tilman and Downing (1994). The higher the number of species, the higher the drought buffer capacity, although the gain per additional plant species decreases with the number of species.

The maintenance of a high biodiversity in all ecosystems has several important reasons:

1. Mankind is using extensively a number of ecosystem services (Costanza et al., 1997), which have a direct and indirect economic value. The services are dependent on a sustainable development of the ecosystems, which according to the discussion presented above would be less probable by a reduced biodiversity. There is a direct economic interest to maintain a high species diversity.
2. A high biodiversity has an aesthetical value, which can best be illustrated by our protection of natural parks and sanctuaries.
3. Biodiversity is a genetic resource for agricultural, pharmaceutical, biotechnological, and chemical industries.

Example 10.3

The species diversity is low toward the poles and high toward the tropics. Give at least a partial explanation of this gradient in the species diversity.

Solution

A detailed and full explanation would require inclusion of many factors. Two factors seem to be particularly important:

1. The productivity gradient associated with the latitude is very similar to the species gradient as a function of the latitude.
2. The development of species diversity, as all biological processes, generally increases with increasing temperature.

10.12 Diversity and Extreme Environment

We have often been surprised by the wide spectrum on the earth of extreme environments that are carrying life. How can life exist several kilometers under the surface of the sea at high concentrations of sulfide and very hot water? The organisms have no light and are exposed to enormous pressure. How can life at all survive in tidal ecosystems with the changing conditions from no water to complete coverage of water? How can life exist in deep caves with complete darkness? How can multicell animals live without oxygen? The reason could be that life will always be present where there are conditions for life, even when the conditions are very harsh. If life has sufficient time to develop, it will inevitably develop, even under harsh or extreme conditions (constraints), provided that some basic requirements or needs are fulfilled (see Section 5.6).

We can, however, not exclude that the possibilities to have life under extreme conditions are determined by the rich differentiation of organisms that we can select from on the earth. Due to the high diversity of species on the earth, there is a high probability that one or more of the species can modify their properties to deal with the conditions in an extreme environment, particularly if sufficient time is available. These considerations are illustrated in Figure 10.6. In other words, the reason we have life in these extreme environments on earth could be the richness of the many different life forms that already have been established. It is also in accordance with the exponential increase of the species diversity, which we have presented in Sections 6.4 and 7.5. The first life on earth was probably formed in very gentle environments—or maybe it came from outer space, and later it was possible with the enormous species diversity to find solutions to how life could evolve under extreme conditions.

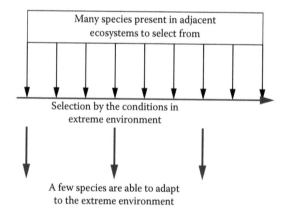

FIGURE 10.6 Life at harsh conditions is only possible if life has had sufficient time to form many different life forms and have many different life forms in adjacent ecosystems, among which some can be modified to meet harsh life conditions.

We can, however, not conclude which of the following two possibilities would give the most correct explanation for life in extreme environments:

1. Life will inevitably emerge everywhere when the conditions are suitable for life, under harsh or gentle conditions.
2. Life at harsh conditions is only possible if life has had sufficient time to form many different life forms, among which some can be modified to meet the harsh life conditions.

Possibility 2 requires a huge diversity. We are mostly inclined to accept the explanation in Figure 10.3 in the light of exponential growth of the species diversity. Later, if we find life in our solar system under harsh conditions, we are closer to giving a proper explanation for the life that we can find on earth in the most extreme environment. It can also not be excluded that you have to use a combination of both explanations.

More details are given below for two characteristic extreme environments to elucidate the points of discussion. The examples are taken from Jørgensen (2008).

10.12.1 The Deep Sea

Until recently the deep sea was considered to be ecosystems with a low abundance of organisms due to the extreme limitation of the food supply. There is, however, one habitat in the deep sea where the density of life is almost equal to what is found in many other marine ecosystems. It is the system of hydrothermal vents, or deep sea hot springs. The vents are found along the ridges at the bottom of the ocean where the earth's crystal parts are spreading apart (Childress et al., 1987). They consist of densely packed animals living several kilometers below the surface. They are living in total darkness, but you can find giant tube worms, for instance, *Riftia pachyptila*, as much as 1 m long, large white clams 30 cm in length (*Calyptogena magnifica*), and clusters of mussels, mainly *Bathymodiolus thermophilus*, forming thick aggregations around the hydrothermal vents. Also shrimps, crabs, and fishes can be found. Photosynthesis is impossible at the depth of the vents. What is the source of energy, because an energy source is absolutely necessary for all ecosystems? Could the density of life be explained by the temperature? The vent water is 10–20°, in contrast to most deep sea water, which is very cold, namely, 2–4°. The warm temperature does, however, not explain the unique life of the vent ecosystems. The explanation is that these underwater springs are rich in hydrogen sulfide, as are many springs on land. Some of the vents are denoted black smokers, because they are colored completely black by the presence of metal sulfide. The sulfide-rich habitats support a large number of bacteria. They are autotrophs not using the sun as an energy source, but the oxidation of hydrogen sulfide. The sulfur bacteria have taken over the job of the green plants closer to the surface. Whereas green plants are phototrophs, the sulfur-oxidizing bacteria are chemotrophs, i.e., capable of using inorganic energy sources to drive carbon dioxide fixation. Figure 10.7 shows a typical and possible food chain close to a deep sea vent (Calaco et al., 2007).

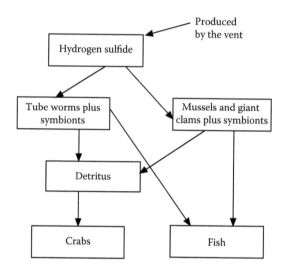

FIGURE 10.7 The food chain in the deep sea at hydrothermal vents. The energy source is hydrogen sulfide, which is bacterial oxidized. The tube worms utilize the chemical energy of hydrogen sulfide by use of endosymbiotic bacteria.

The hydrothermal vent communities consist of unusual invertebrates species: a dense cluster of tube worms, vent mussels, and vent crabs. The tube worm is essentially a closed sac without a mouth and digestive system. At its anterior tip there is a red gill-like plume where oxygen, carbon dioxide, and hydrogen sulfide are exchanged with the ambient seawater. The animal consists mainly of a thin-walled sac that contains the internal organs. The largest of them is the trophosome, which occupies most of the body cavity. It contributes significantly to the worm's nutrition and is colonized by a vast number of sulfur-oxidizing bacteria. The tube worm and the bacteria have established what is known as an endosymbiotic relation. The tube worm receives reduced carbon molecules from the bacteria, and in return provides the bacteria with the raw materials needed to fuel its metabolism: carbon dioxide, oxygen, and hydrogen sulfide.

The crucial question is how the fauna of the vent can survive the high concentration of the toxic hydrogen sulfide, which is able to block the respiration. Studies of the tube worm have revealed that sulfide has no effect on oxygen binding, and that the worm's respiration is substantial even in the presence of sulfide concentrations lethal to most animals. The tube worms can extract sulfide from vent water, and they have a blood component, the worm's high molecular weight hemoglobin, that is able to bind hydrogen sulfide stronger than the sulfide-sensitive cytochrome c oxidase. The hemoglobin can bind oxygen and hydrogen sulfide simultaneously but at different sites. The large white clam and the mussels have also developed a symbiotic relation to the sulfur bacteria. The bacteria are not, however, in the internal organ, but in the gills, where they can readily obtain oxygen and carbon dioxide from the respiratory water flow. The metabolic plan is, however, the same, namely, that the bacteria oxidize sulfide and supply the clam with fixed carbon compounds. The clams concentrate the sulfide in their blood. The level is orders of magnitude higher than the concentrations in the ambient water. Apparently, the clams absorb sulfide through their large elongated feet, which extend into the hydro-thermal vents, where the concentrations of sulfide are highest. After adsorption through the clam's feet, sulfide is transported by the blood to the bacteria in the gills. The trans-port takes place by a special high molecular protein, which is able to protect the sulfide against oxidation on the way to the gills and protect the hemoglobin and cytochrome c oxidase. The binding of sulfide to the protein is reversible. It is offloaded to the bacteria in the gills and oxidized to provide energy.

The tube worm, the clam, and the mussels owe their ecological success to their sym-biosis with sulfur bacteria (Calaco et al., 2007). Many of the smaller vent animals lack symbionts. They obtain their nutrients either by filtering particular food such as bacte-ria from the water or by feeding on animals that do contain symbionts. Vent crabs, for instance, feed on the respiratory plume of tube worms. Many of the animals in the deep sea can produce light by chemical reactions of special proteins, for instance, fish, octo-pus, and squid. There are also light-producing bacteria. Clearly it is a great advantage to produce light; otherwise there is complete darkness. The hydrothermic activity in the deep sea is not only the energy source directly or indirectly for many life forms, but also an important source of minerals for the entire marine life.

10.12.2 Carnivorous Plants

The luring, capturing, and digesting mechanisms that have evolved in some plants enable them to capture insects in order to augment their supply of nutrients and thereby sur-vive in habitats where few other plants can live. The carnivorous plants can be divided into two groups according to their methods of catching prey. They are either active or passive trappers (Heslop-Harrison, 1978). The active trappers catch hopping or crawl-ing insects. When the prey are touching the leaf, tactile hairs are agitated and a closing mechanism is triggered.

For the Venus flytrap, the two sides of the leaf quickly move together, closing the trap.

For the bladderwort, the flap of the tissue that forms the door swings open, and the bladder expands suddenly to draw in both water and prey.

The passive trappers, for instance, the pitcher plant, lure the prey to a slippery edge and the prey falls into a pool of digestive fluid and cannot climb out. Another passive trapper, the sundew, has attractively colored leaves. When small flying insects touch the secretion globules on the leaf surface, digestive enzymes are secreted. The nutrients derived from captured prey enter the leaves at a surprisingly high rate.

It has been shown by numerous experiments that the carnivorous plants performed better when provided with prey. The main function of the carnivorous habit is to pro-vide scarce nutrients. It is therefore expected that they inhabit environments where such supplementation would be beneficial, which is also in accordance with observations. The carnivorous plants are encountered most frequently in nutrient-poor ecosystems, where the constraints are too low concentrations of the nutrients' heats, bogs, impoverished soil in forest openings, and on clay soil associated with weathered limestone.

The carnivorous plants occupy extremely narrow ecological niches. The range of adaptations found in the carnivorous plants is furthermore completely according to Darwin's view as examples of evolutionary virtuosity. Four hundred out of the about

300,000 species of flowering plants are known to be carnivorous. They belong to 13 or so genera and 6 families.

The supplemental nutrients available to the carnivorous plants from the digestion of the prey offer them special advantages, particularly in environments where nutrients are scarce. Both nitrogen and phosphorus seem to play an important role in this context. But are these advantages counterbalancing the costs? The plants have to invest energy in the synthesis of digestive enzymes and other secretion products and in the plant's elaborate structural adaptation. The carnivorous plants are, however, found in places with adequate carbon sources, access to water, and no limitations of the photosynthetic energetic resource. So, the question is not about the energy cost, but about survival in places where no noncarnivorous competitors can intrude. Whatever the energy cost may be, the investment is justified by the survival possibilities.

In this context it should be mentioned that the differentiation goes so far that there are also animals that use photosynthesis to obtain extra energy. The naked marine snail *Elysia chlorotica* has the shape of a leaf. The snail has taken up genes from algae that makes it possible for the snail to convert carbon dioxide to glucose. The snail produces proteins that are able to maintain chlorophyll-a in its cells. *Elysia chlorotica* illustrates that genes can be exchanged among species, which has increased the differentiation.

10.13 Summary of Important Points in Chapter 10

1. The combinations of forcing functions (constraints) on ecosystems vary enormously with time and space. The evolution has been almost 4 billion years, which implies that the ecosystems have had a very long time to find many possible solutions to the challenge to grow and develop. The ecosystems have therefore developed an enormous diversity in all levels of the ecological hierarchy: the biochemical molecules, the cells, the organs, the individuals, the species, the populations, the communities, the ecosystems, and the entire ecosphere.

2. High diversity implies a wider spectrum of ecosystem services. There is not a simple relationship between diversity on the one side and resistance, buffer capacity of resilience on the other side. High diversity implies, however, that the spectrum of probability to meet and cope with forcing functions (eventually new and unexpected forcing functions) increases.

3. Even an extreme environment has life, probably as a consequence of the high diversity on all the hierarchical levels.

Exercises/Problems

1. You share 98% of your genes with a chimpanzee and 99% with other humans. What percentage of your genes do you share with your sister?

2. List the advantages of the hierarchical organization that is presented in Chapter 9, and comment on these advantages in the light of the diversity on each level, as presented in this chapter.

3. An ecosystem consists of about 1 million individuals and 500 species. The 20 most dominant species have about 20,000 individuals each, while the 20 second most

abundant species have about 10,000 individuals. The rest of the species are covered by the remaining 400,000 individuals, with approximately the same number of individuals for each of the 460 species. Find Shannon's index and the index of evenness.

4. Why should we expect that diversity has increased over time (see also Section 7.5)?

5. List the advantages of a high species diversity.

6. There is diversity on each level of the hierarchy, but why is species diversity of particular importance?

11

Ecosystems Have a High Buffer Capacity

Believe nothing, no matter where you read it
Or who said it
No matter if I have said it
Unless it agrees with your own reason
And your own common sense.

Buddha

The stability concepts resilience, resistance, and buffer capacities are defined and illustrated with examples from ecosystems. The two latter can easily be quantified and are multidimensional, which is necessary to describe reactions of such complex systems as ecosystems.

A high sum of buffer capacities complies with a high eco-exergy according to models and statistics. A high biodiversity ensures a wider spectrum of buffer capacities.

The intermediate disturbance hypothesis and hysteresis behavior of ecosystems are presented, discussed, and illustrated.

Ecosystems can behave chaotically. They strive toward a high utilization of the resources (this means to get as much eco-exergy as possible—compare with ELT), but this means that they also may operate at the edge of chaos; however, they have much feedback to avoid the chaos, which could give a high probability for less eco-exergy on a long-term basis.

11.1 Introduction: Stability Concepts

Are ecosystems generally stable? What makes them unstable? These two questions are among the core problems of system ecology, but we can easily lengthen the list of relevant questions: Can ecosystems persist in the course of a sufficiently long time in spite of perturbations, changing forcing functions, constraints, or impacts? What will happen when the perturbations stop? Or the forcing functions return to normal values valid before the major impacts or disturbances started? We will try to answer these questions in this chapter, but before starting the discussion of the questions, it is necessary to clarify what we mean by stability and give definitions of the most important stability concepts. Though the stability concepts have been widely discussed in the ecological literature, it is difficult to provide a precise and unambiguous definition. For those seeking a mathematical angle to the stability concept, refer to Chapter 6 in Jørgensen and Svirezhev (2004).

Three stability concepts are defined, applied, and discussed in this chapter:

Resilience is generally in science defined as the capability of a strained body to recover its size and shape after deformation caused especially by compressive stress. It has been introduced in ecology by Holling (1986) as "the capacity of a system to absorb disturbances and reorganize while undergoing change so as to still retain essentially the same function, structure, identity, and feedbacks," but a slightly more quantitative definition can also be found in the ecological literature: the maximum amount a system can be changed before losing its ability to recover. Some ecological systems display, however, several possible stable states. They may also show a hysteresis effect in which, even after a long time, the state of the system may be partly determined by its history. The concept of resilience depends therefore upon our objectives, the types of disturbances that we anticipate, control measures that are available and desirable, and the timescale of interest. Ecosystems are very complex, and this makes it difficult to apply the concept of resilience, at least quantitatively. It is necessary, if the concept should be applied in environmental management, to define what is the exact meaning of "the same function and structure" and which disturbances are actual. An ecosystem will never return to the same conditions and state again (see, for instance, the variability of constraints in Section 10.2). If specific disturbances are examined and the core functions and the most important elements of the structure all are clearly defined, it makes sense to use the concept in environmental management. Otherwise, the concept can only be applied vaguely in a qualitative context.

Resistance is understood as the ability to resist changes, when the impact on the ecosystem or its forcing functions are changed or perturbations are introduced. It can easily be applied quantitatively as the ratio of changes in the ecosystem of forcing functions to the changes of state variables or process rates. The concept could be used multidimensionally in the sense that there are many different ratios that could be used corresponding to all relevant combinations of forcing functions and state variables. The stability information that we gain by the use of this concept is more quantitative than resilience. On the other hand, resilience looks into the ability of the ecosystem to recover, if the forcing functions are changed more significantly. Therefore, both concepts are relevant in environmental management and ecosystem conservation, and they supplement each other. Resistance gives a good quantitative answer to the stability questions, but still it is relevant to discuss: Can the ecosystem recover—particularly, can it recover its functions and main structure? This question should be answered in as much detail and as quantitatively as possible.

Buffer capacity is used for the title of the chapter to emphasize that it is associated with the quantitative presentation of system ecology; that is the basis for this textbook. However, it is very much related to resistance and has the same immediate advantages and disadvantages.

Buffer capacity, β, is defined as follows (Jørgensen, 1994a, 2002):

$$\beta = 1/(\partial \text{ (state variable)}/\partial \text{ (forcing function))} \tag{11.1}$$

or in environmental management context as Δ forcing function/Δ state variable

Forcing functions are the external variables that drive the system, such as discharge of wastewater, precipitation, wind, and so on, while state variables are the internal variables that determine the system, for instance, the concentration of soluble phosphorus, the concentration of zooplankton, of certain species, and so on. As seen, the concept of buffer capacity has a definition that allows us to quantify, for instance, in modeling, and it is furthermore applicable to real ecosystems, as it acknowledges that *some* changes will always take place in an ecosystem in response to changed forcing functions. The question is how large these changes are relative to changes in the conditions (the external variables or forcing functions). The concept should be applied multidimensionally, as we may consider a number of relevant combinations of state variables and forcing functions. This implies that even for one type of change there are many buffer capacities corresponding to each of the state variables. Rutledge et al. (1976) define ecological stability as the ability of the system to resist changes in the presence of perturbations. It is a definition very close to buffer capacity, but it still lacks the multidimensionality of ecological buffer capacity.

The relation between forcing functions (impacts on the system) and state variables (indicating the conditions of the system) is rarely linear, and buffer capacities are therefore not constant, but strongly dependent on the state of the ecosystem. It may therefore, in environmental management, be important to reveal the relationships between forcing functions and state variables to observe under which conditions buffer capacities are small or large; compare with Figure 11.1.

In Chapter 10 it was mentioned that a high buffer capacity may sometimes at least be accompanied by low diversity (see Figure 10.4). It was furthermore mentioned that a high diversity gives a higher probability for the ecosystem to meet new and sometimes unexpected impacts, and that it may sometimes at least imply higher resistance or buffer capacity (see Figure 10.3). Under all circumstances, the stability of ecosystems and the biodiversity cannot be expressed by a relatively simple analytical equation, but the relationship between these two important ecosystem concepts is very complex (May, 1973)

A statistical analysis of several model studies (Jørgensen, 2002) has shown that there is a strong correlation between the sum of many (relevant) buffer capacities and the eco-exergy of the considered ecosystem. If this is right, eco-exergy can be considered an indicator for the sum of buffer capacities (resistance) of an ecosystem. It emphasizes the importance for ecosystems to gain as much eco-exergy as possible, which they do according to ELT.

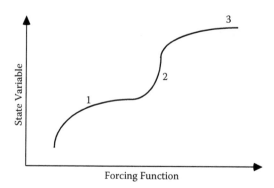

FIGURE 11.1 The relation between state variables and forcing functions is shown. At points 1 and 3 the buffer's capacity is high; at point 2 it is low.

Moreover, it is in accordance with Svirezhev (1990), who showed that the energy needed to dismantle an ecosystem is at least equal to the eco-exergy of the ecosystem.

To summarize, buffer capacity and resistance seem to be applicable stability concepts in environmental management, as they are based:

1. On an acceptance of the ecological complexity—it is a multidimensional concept, as there is a buffer capacity to each combination of forcing functions and state variables.
2. On reality, i.e., that an ecosystem will never return to exactly the same situation again.

The environmental management could be supplemented by such questions as:

- Are there possibilities that the ecosystem will be able to return to a normal function?
- Can the ecosystem recover and offer the ecosystem services that we are dependent on?

This means that it is still relevant to have the concept resilience in mind in environmental management, because ecosystems may have a high buffer capacity and not easily recover after major disturbances, or may have a low buffer capacity and still be able to recover.

Example 11.1

Do the observed buffer capacities for eutrophication and toxic substances comply with ELT and the above-mentioned statistical analysis, which shows that eco-exergy is a good measure of the sum of buffer capacities?

Solution

The answer is yes. A lake that has a high concentration of phytoplankton has a very high buffer capacity for changes of the eutrophication level. Some other buffer capacities may be low, but the eco-exergy is at least relatively high due to the high biomass concentration. When a toxic substance has a high concentration and several species may be extinguished, the buffer capacity is relatively high for changes in concentration of the toxic substances, because the resistant species have survived. The eco-exergy has most probably decreased due to the reduced biomass, but ELT emphasizes that the system tries to obtain the highest possible eco-exergy under the prevailing conditions, which are, of course, less favorable when a toxic substance is discharged to an ecosystem.

11.2 The Intermediate Disturbance Hypothesis (IDH)

Disturbance of ecosystems is defined by Picket and White (1985) as a relatively discrete event in time characterized by a frequency, intensity, and severity outside a predictable range. In a famous paper, Hutchinson (1948, 1957, 1965) describes what he called the plankton paradox: Many more species of phytoplankton have been observed to coexist in a relatively simple environment. He presumed that the species richness observed was a consequence of fluctuations in the environment, which prevent it from attaining a steady state. So, the coexistence of many species is the result of a nonequilibrium phenomenon. Disturbances may work in favor of species diversity by lowering the pressure of dominant species upon other species and allowing the latter to develop. Based upon many ecological investigations, Cornell (1978) proposed the intermediate disturbance hypothesis: species richness will be greater in communities with moderate levels of perturbations than in communities without disturbances, or communities with rare or very frequent disturbances (Figure 11.2). He used the hypothesis to explain the high diversity found in tropical forest ecosystems and coral reefs.

A structurally dynamic model was applied by Jørgensen and Padisák (1996) to understand the seasonal changes of biodiversity in Lake Balaton. They found that intermediate disturbances could explain the high biodiversity in Lake Balaton from April 1 to late October (except in early July, where there were no changes and the biodiversity was low). The high biodiversity also coincides with a high eco-exergy (except in July, where it was significantly lower). The model studies showed clearly how intermediate disturbances were working: by giving the opportunity to many species (mainly phytoplankton species) to grow under different conditions that were a consequence of the disturbance. Only a few species were extinct, because the unfavorable conditions were changed before the species died out. When the conditions were relatively constant for a long time, a few species were favored and the other species became extinct. The model results can be summarized as follows:

Intermediate disturbances favor species diversity and give a high eco-exergy and sum of various relevant buffer capacities, while constant conditions give a low diversity and a relatively low eco-exergy and a low sum of buffer capacities.

11.3 Hysteresis and Buffer Capacities

A lake ecosystem even at low levels of eutrophication shows a remarkable buffering capacity by an increasing nutrient level (Jørgensen, 1990), which can be explained by a current increasing removal rate of phytoplankton by grazing and settling. Zooplankton and

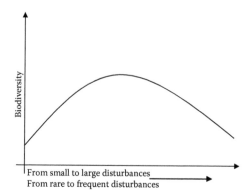

FIGURE 11.2   The graph shows biodiversity as a function of the magnitude and frequency of the disturbances, according to Cornell (1978).

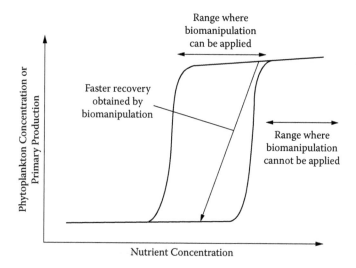

FIGURE 11.3   The hysteresis relation between nutrient level and eutrophication measured by the phytoplankton concentration is shown. The possible effect of biomanipulation is shown. An effect of biomanipulation can hardly be expected above a certain concentration of nutrients, as indicated in the diagram. The biomanipulation can only give the expected results in the range where two different structures are possible.

carnivorous fish abundance are maintained at relatively high levels under these circumstances. At a certain level of eutrophication it is, however, not possible for zooplankton to increase the grazing rate further. The grazing rate can be described by a Michaelis-Menten equation (see Figures 2.2 and 2.3). When the grazing rate has reached the maximum corresponding to a high concentration of phytoplankton, it becomes independent of the phytoplankton concentration. The phytoplankton concentration may increase very rapidly under these circumstances by even slightly increasing concentrations of nutrients. The eutrophication and remediation of a lacustrine environment do not proceed according to a linear relationship between nutrient load and vegetative biomass, but rather display a sigmoid trend with delay, as shown in Figure 11.3. The hysteresis reaction is completely in accordance with observations (Hosper, 1989; Van Donk et al., 1989), and it can be explained by structural changes (de Bernardi, 1989; Hosper, 1989; Sas, 1989; de Bernardi and Giussani, 1995). When the nutrient input is decreasing under these conditions a similar buffering capacity to variation is observed. The structure has now changed to a high concentration of phytoplankton and planktivorous fish, which causes a resistance and delay to a change where the second and fourth trophic levels become dominant again. The described structurally dynamic behavior has been modeled (see Jørgensen and de Bernardi, 1997, 1998).

Willemsen (1980) distinguishes two possible conditions:

1. A bream state, characterized by turbid water, high eutrophication, low zooplankton concentration, absence of submerged vegetation, large amount of breams, while pike, which is a top carnivorous fish, is hardly found at all.
2. A pike state, characterized by clear water and low eutrophication. Pike and zooplankton are abundant, and there are significantly fewer breams.

In the late 1970s, biomanipulation was proposed as a method to recover lakes. The idea was to change the conditions from a bream state and high eutrophication to a pike state and low eutrophication by removal of the breams (see Figure 11.4) and introduction of the pike. The method was tested particularly in Italy, the Netherlands, and Denmark, and sometimes it was successful and sometimes it failed. The presence of two possible states in a certain range of nutrient concentrations may explain why biomanipulation has not always been used successfully.

FIGURE 11.4 (See color insert.) Removal of planktivorous fish by a fishery in a North Italian lake.

According to the observations referred to in the literature, success is associated with a total phosphorus concentration below 120–140 μg/L (Jeppesen et al., 1990), while disappointing results are often associated with a phosphorus concentration above this level (Benndorf, 1987, 1990), with difficult control of the standing stocks of planktivorous fish (Shapiro, 1990; Koschel et al., 1993).

Scheffer (1990) has used a mathematical model based on the catastrophe theory to describe these shifts in structure. However, this model does not consider the shifts in species composition, which is of particular importance for biomanipulation. The zooplankton population undergoes a structural change when we increase the concentration of nutrients, for example, from a dominance of calanoid copepods to small caldocera and rotifers, according to the following references: de Bernardi and Giussani (1995) and Giussani and Galanti (1995).

Hence, a test of structurally dynamic models could be used to give a better understanding of the relationship between concentrations of nutrients and the vegetative biomass, and to explain possible results of biomanipulation. The test has been successful (see Jørgensen and Fath, 2011) and made it possible to understand the above-described changes in structure and species compositions. The details of the model are published in Jørgensen and de Bernardi (1998). The applied model has six state variables: (1) dissolved inorganic phosphorus; (2) phytoplankton, phyt.; (3) zooplankton, zoopl.; (4) planktivorous fish, fish 1; (5) predatory fish, fish 2; and (6) detritus. The forcing functions are the input of phosphorus, in P, and the throughflow of water determining the retention time. The latter forcing function also determines the outflow of detritus and phytoplankton.

The model results follow closely the hysteresis shown in Figure 11.3 and the structural changes, from (a) dominance of zooplankton and carnivorous fish to (b) dominance of phytoplankton and planktivorous fish, take place at the expected concentrations of phosphorus, namely, at 0.12–0.13 and 0.06–0.065 mg/L, respectively, provided that P is the limiting nutrient, which is most often the case for lakes. The eco-exergy is considerably higher by the (a) structure below 0.06 mg P/L and above 0.13 mg P/L for the (b) structure, while the two possible structures have approximately the same eco-exergy between these two concentrations. There are thus two possible structures that can give the highest eco-exergy, and this explains the hysteresis behavior.

Example 11.2

A lake receiving wastewater with a phosphorus concentration of 3 mg/L is eutrophic. The wastewater is diluted by natural streams with no phosphorus 10 times, and the phosphorus concentration in the lake is 0.3 mg/L, corresponding approximately to the 10 times dilution. The water retention time in the lake is 3 years. It is possible with a relatively smaller investment to reduce the phosphorus concentration by chemical precipitation in the wastewater to 0.8–1 mg/L, while a reduction to a lower concentration is possible but would require a more massive investment.

Would biomanipulation give an acceptable reduction of the eutrophication? It could be a beneficial method to use because the application of the method has moderate costs.

Which method or combination of methods would you propose to use for an acceptable reduction of the eutrophication?

Solution

Biomanipulation cannot be applied solely because the phosphorus concentration is above 0.13 mg/L. A treatment of the wastewater by the cost-moderate method would reduce the phosphorus concentration to about 0.08–0.1 mg/L, which is not sufficient to cause a change in the structure.

A combination of the cost-moderate waste treatment and biomanipulation would solve the problem, because the phosphorus concentration would, during a period of probably 6–9 years, corresponding to two to three times the water retention time, easily be reduced to 0.08–0.1 mg/L, which would ensure that the biomanipulation would change the structure.

Scheffer and coworkers (2001) have found a similar hysteresis behavior of shallow lakes.

Below about 0.1 mg P/L submerged vegetation is dominant, and above about 0.25 mg P/L phytoplankton is dominant and is able to outcompete the submerged vegetation by shading. Between 0.1 and 0.25 mg P/L both submerged vegetation and phytoplankton can be dominant, depending on the history.

Submerged vegetation has in shallow lakes high eco-exergy and buffer capacities, which means that phytoplankton cannot, by increasing phosphorus concentration, outcompete the submerged vegetation before the concentration has reached 0.25 mg P/L, while submerged vegetation cannot outcompete phytoplankton by a decreasing phosphorus concentration before the concentration has reached 0.1 mg P/L. There are therefore two solutions to the highest possible eco-exergy between 0.1 and 0.25 mg P/L.

Zhang et al. (2003a, 2003b) have developed a structurally dynamic model by use of STELLA to show the hysteresis behavior in the case of shallow lakes, with competition between submerged vegetation and phytoplankton. The conceptual diagram of the model is shown in Figure 11.5. The model was developed by use of data from Lake Mogan near Ankara, Turkey. Phosphorus is the limiting factor for eutrophication in the lake, which is interesting to apply for this examination, because it is a shallow lake with a competition between phytoplankton and submerged vegetation. The water retention time in the lake is, on average, almost 1 year. The model has seven state variables: soluble P, denoted PS; phosphorus in phytoplankton, PA; phosphorus in zooplankton, PZ; phosphorus in detritus, PD; phosphorus in submerged plants, PSM; exchangeable phosphorus in the sediment, PEX; and phosphorus in pore water, PP. The processes are inflows and outflows of phosphorus, phosphorus in phytoplankton,

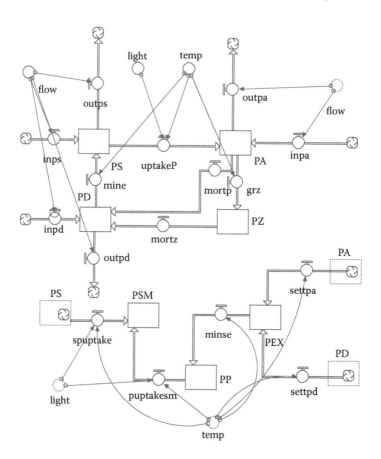

FIGURE 11.5 (See color insert.) The conceptual diagram of the Lake Mogan eutrophication model focusing on the cycling of phosphorus. The model has seven state variables: soluble P, denoted PS; phosphorus in phytoplankton, PA; phosphorus in zooplankton, PZ; phosphorus in detritus, PD; phosphorus in submerged plants, PSM; exchangeable phosphorus in the sediment, PEX; and phosphorus in pore water, PP.

and phosphorus in detritus. Soluble phosphorus is taken up by phytoplankton, and the process is named uptakeP. Zooplankton grazes on phytoplankton, indicated grz on the diagram. The settling of detritus and phytoplankton is covered by a first-order reaction. A part of the settled material is lost as nonexchangeable phosphorus, while the exchangeable fraction goes to the state variable exchangeable phosphorus, PEX. A mineralization of the exchangeable phosphorus takes place in the sediment, and the process is named minse in the diagram. Mineralization in the water phase of detritus phosphorus is a process named mine. Temperature influences all the process rates. Light is considered a climatic forcing function influencing the growth of both phytoplankton and submerged plants. The submerged plants take most phosphorus up from the sediment, but the model considers both uptake of phosphorus from water and sediment by the submerged plants.

In accordance with the procedure for development of SDMs (see Figure 7.4), the model should be tested for all combinations of at least three possible values of the variable parameters. This means that if seven parameters are made structurally dynamic, as was decided for the eutrophication model in this case, it is required to run the model 3^7 times = 2,187. To minimize the number of testable combinations, Zhang et al. used allometric principles. Eco-exergy was calculated for the model as eco-exergy = 21*phytoplankton + 135*zooplankton + 100*submerged vegetation + detritus. The optimization of the eco-exergy was repeated every 10 days. After an annual model run, it was possible to give a graph or table of the phytoplankton and zooplankton sizes, that by use of the applied allometric principles would be "translated" to the seven parameters that were selected for the development of the structurally dynamic model.

The structurally dynamic model was used after calibration to answer the following question: Is it possible to describe the structural changes from submerged vegetation dominance to phytoplankton dominance and back again, as described by Scheffer et al. (2001)? Can we simulate the described hysteresis behavior by use of the developed SDM for Lake Mogan? The phosphorus concentration in the lake is about 80–85 µg/L according to the observations, and it was the concentration applied for the calibration described above. To answer the questions, the phosphorus concentration in the water was increased by a factor of 5 times, which implies that we should expect a phosphorus concentration after a couple of years of about 400 µg/L. Afterwards the phosphorus concentration was reduced to the present level of 80–85 µg/L. Figure 11.6 illustrates the results of these changes in the phosphorus concentration from 80–85 to 400 µg/L and from 400 back to 80–85 µg/L.

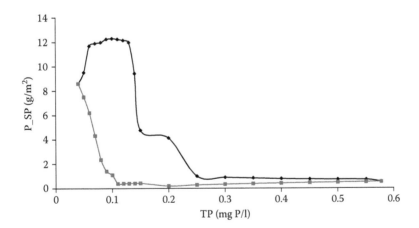

FIGURE 11.6   (See color insert.) The graphs show the reaction of submerged plant phosphorus to an increase of phosphorus to about 400 µg/L (black curve), followed by return to the original, about 80–85 µg/L (grey curve). When the phosphorus concentration is increased, P-SP increases due to the higher concentration of nutrients, but at about 250 µg/L the submerged vegetation disappears and is replaced by phytoplankton-P. At the return to 80–85 µg/L the submerged vegetation emerged at 100 µg/L. The hysteresis behavior is completely in accordance with Scheffer et al. (2001).

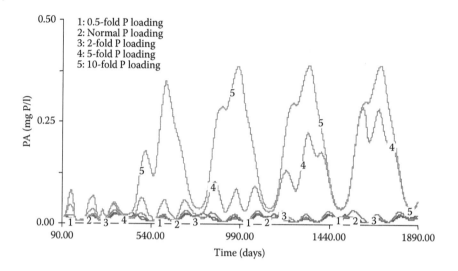

FIGURE 11.7   (See color insert.) The phytoplankton-P concentration as a function of time is shown when the phosphorus concentration of the water flowing into the lake is increased by (1) a factor of 0.5, (2) normal (factor of 1.0), (3) a factor of 2, (4) a factor of 5, and (5) a factor of 10. The present concentration of phosphorus in the lake is 80–85 µg/L. (1) to (3) give no changes, while (4) and (5) give a significant increase of phytoplankton, which becomes dominant.

Figure 11.7 shows the phytoplankton-P concentrations as a function of time for five different phosphorus concentrations of the water flowing into the lake: (1) 0.5 times the present level, (2) the present level, (3) 2 times the present level, (4) 5 times the present level, and (5) 10 times the present level. The present level is about 80–85 µg P/L.

Figure 11.8 shows the submerged plant as g P/m² as a function of time for the same five phosphorus concentrations of the water flowing to the lake. The shift in dominance from submerged vegetation to phytoplankton is very clear for the phosphorus concentrations 5 and 10 times the present value.

The two presented cases of hysteresis behavior of lakes have shown that ecosystems may behave in a complex way to changed forcing functions, but that it is possible to explain the behavior by the use of ELT: ecosystems strive toward highest possible eco-exergy = work capacity, which also can explain the shifts of structural changes and of the buffer capacity.

Submerged vegetation is preferable because it implies clear water, and therefore also denotes the clear water structure or conditions (see Figure 11.9).

11.4 Chaos, Disturbances, and Buffer Capacities

Kauffman (1993, 1995) has hypothesized the proposition that biological systems operate at the edge of chaos. He studied a Boolean network and found that a network on the boundary between order and chaos may have the flexibility to adapt rapidly and

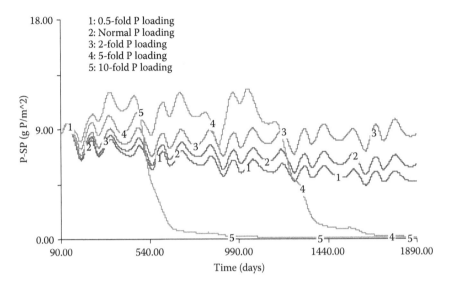

FIGURE 11.8 (See color insert.) The submerged plant concentration as gP/m² as a function of time is shown when the phosphorus concentration of the water flowing into the lake is increased by (1) a factor of 0.5, (2) normal (factor of 1.0), (3) a factor of 2, (4) a factor of 5, and (5) a factor of 10. The present concentration of phosphorus in the lake is 80–85 μg/L. (1) to (3) give no changes, while (4) and (5) first give a minor increase of the concentration, followed by a significant decrease, when phytoplankton becomes dominant; compare with Figure 11.6.

FIGURE 11.9 (See color insert.) Clear water stage or structure by dominance of submerged vegetation.

successfully through the accumulation of useful variations. In such poised systems most mutations will have small consequences because of the system's homeostatic nature. Such poised systems will typically adapt to a changing environment gradually,

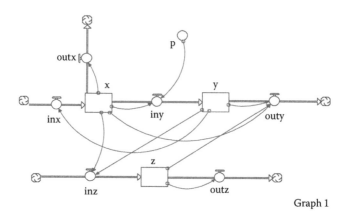

Graph 1

FIGURE 11.10   (See color insert.) A simple model that shows chaotic behavior.

but if necessary, they can occasionally change rapidly—properties that can be found in organisms and ecosystems. This explains, according to Kauffman, why Boolean networks poised between order and chaos can generally adapt most readily, and therefore have been the target of natural selection. Chaos theory is concerned with unpredictable courses of events. The irregular and unpredictable time evolution of many nonlinear systems has been named chaos.

> Chaos theory has eliminated the Laplacian illusion of deterministic predictability and can therefore be conceived as a ticking bomb under reductionistic science.

Even very simple models can behave chaotically. The very simple model shown in Figure 11.10 with equations in Table 11.1 behaves chaotically at certain values of the parameter, p. The model could represent an ecological system or subsystem with several feedback controls: three populations x, y, and z with a significant control of each other. For p = 25 (see Figure 11.11), the model will easily reach steady state, but if p = 33, for instance, the model behaves chaotically. When such a simple model behaves very differently for a minor change in a parameter, how can we develop models of very complex biological systems?

Chaos theory is best illustrated by Lorentz's (1955, 1963) famous butterfly effect—the notion that a butterfly stirring the air in Hong Kong today can transform storm systems in New York next month. The effect was discovered accidentally by Lorentz in 1961. He was making a weather forecast and wanted to examine one sequence of greater length. He tried to make what he thought was a shortcut. Instead of starting the whole run over again, he started halfway through. To give the computer its initial values, he typed the numbers from the earlier printout. The new run should therefore duplicate the old one, but it did not. Lorentz saw that his new weather forecast was diverging so rapidly from the previous run that within a few months all resemblance had disappeared. There had been no malfunction of the computer or the program. The problem lay in the number he had typed. In the computer six decimal places were stored: 0.506127, but to save time, because he thought it was unessential, he printed a rounded-off number with just three

TABLE 11.1 Equations for the Model Shown in Figure 11.8

$x(t) = x(t - dt) + (inx - iny - outx)*dt$
INIT x = 1

Inflows

inx = 10*y

Outflows

iny = p*x
outx = 10*x
$y(t) = y(t - dt) + (iny - outy)*dt$
INIT y = 1

Inflows

iny = p*x

Outflows

outy = y + x*z
$z(t) = z(t - dt) + (inz - outz)*dt$
INIT z = 1

Inflows

inz = x*y

Outflows

outz = 8*z/3
p = 20

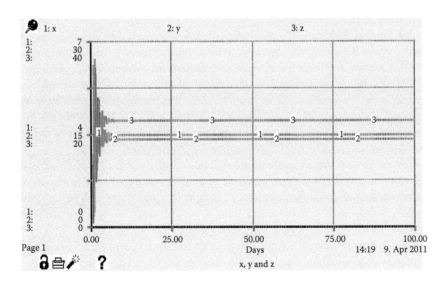

FIGURE 11.11 (See color insert.) Simulations of the model in Figure 11.8, using iny = 25*x (see the equations in Table 11.1).

decimals: 0.506. The explanation is simple: Lorentz's model is very sensitive to initial conditions, and so is the weather itself, and ecological and biological systems, too. The effect is observed today in numerous relations, and all ecological modelers know this problem. Therefore the initial values of the state variables are most often included in a modeler's sensitivity analysis.

The definition of deterministic chaos implies that the distance between the two curves with slightly different initial conditions is growing exponentially:

$$d(t) = d(0)*e^{L*t} \tag{11.2}$$

where $d(t)$ is the distance at time = t, $d(0)$ is the distance at time = 0, and L is a positive number, called the Lyapunov exponent, which is a quantitative indicator for chaos.

After the time $1/L$ the initial conditions are insignificant, i.e., "forgotten." The Lyapunov exponent can be found by plotting the logarithm of the distance between the two curves, neglecting the distance at time 0 (which is 0) vs. the time.

Chaos is also known in relation to bifurcation, and this form of chaos is nicely illustrated by examination of a simple model in population biology. May (1973) has examined the behavior for nonlinear differential and difference equations, for instance:

$$N_{t+1} = N_t (1 + r(1 - N_t/K)) \tag{11.3}$$

where N is the number of individuals in the population under consideration, r is the growth rate per capita, t is the time, and K is the carrying capacity of the environment. Notice that this equation expresses a time delay = 1 in the form in which the difference equation is given. As long as the nonlinearity is not too severe, the time delay built into the structure of the difference equation (11.3) tends to compare to the natural response time of the system, and there is simply a stable equilibrium point at $N\# = K$. However, for $r \geq 2$ this point becomes unstable. It bifurcates to produce two new and locally stable fixed points of period 2, between which the population oscillates stably in a 2-point cycle. With increasing r, these two points also bifurcate to give four stable fixed points of period 4. In this way, through successive bifurcations, an infinite hierarchy of stable cycles of period 2n arises. Figure 11.12 illustrates the formation of bifurcations up to r = 2.75.

When we consider the many nonlinear relationships that are valid in ecology, we may wonder why chaos is not observed more frequently in nature, or even in our models. An obvious answer could be that nature attempts to avoid chaos, and as opposed to the physical system, the ecosystem has many possible, hierarchically organized regulation mechanisms to avoid chaotic situations (see Chapter 9). This does not imply that chaotic or "almost chaotic" situations are not observed in ecosystems. They are, only more rarely observed than expected from the complexity and feedback regulations of ecosystems. The classical example is the almost legendary lemming (Shelford, 1943). According to this paper r*T is 2.4, with r being the growth rate per capita and T the time lag. Oscillation between two steady states should be expected, as Shelford also found

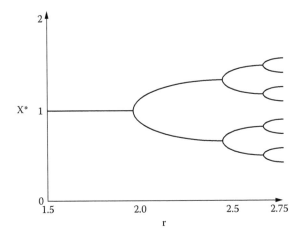

FIGURE 11.12 The hierarchy of stable fixed points of periods 1, 2, 4, 8, ..., 2n, which are produced from Equation (11.3) as the parameter r increases. The y-axis indicates relative values.

(Shelford, 1943). Hassel et al. (1976) culled data on 28 different populations of seasonal breeding insects. They found that the growth may be described by a difference equation as follows:

$$N_{t+1} = q^*N_t \, (1 - a^*N_t)^{-\beta} \tag{11.4}$$

q here is related to r as follows: $r = \ln q$. a and β are constants.

Figure 11.13 shows the theoretical domains of stability behavior for Equation (11.4) applied to the 28 populations by Hassel et al. By far most of the populations are in the monotonic damping area, and only one is in the chaos area (and, as indicated by Hassel et al., it is a laboratory population) and one in the stable limit cycles area. Notice that there is a tendency for laboratory populations to exhibit cyclic and chaotic behavior, whereas natural populations tend to have a stable equilibrium point. The laboratory populations are maintained in a homogeneous environment and are free from predators and many other natural mortality factors, that up to a certain level may very well give a stabilizing effect due to a number of feedback mechanisms.

The relationship between the parameters and the somewhat chaotic behavior is discussed below. It may be concluded that natural populations are able to avoid chaotic situations to a large extent. The long experience gained during evolution has taught the natural population to omit the properties, i.e., the parameters, that may give chaotic situations, because they simply threaten their survival, at least in some situations. Furthermore, the natural populations have the flexibility to select a combination of parameters that give a better chance for survival, because they give a higher eco-exergy, and therefore survival according to ELT, and also a high total sum of buffer capacities.

Figure 11.14 shows a model that has been applied for modeling experiments. We have, however, excluded fish as a state variable in the first place. The phytoplankton and the bacteria were given the maximum growth rates found in the literature. We have then asked: Which are the right maximum growth rates for the two zooplankton state variables, and can we, by a proper selection of the maximum growth rates, avoid chaotic

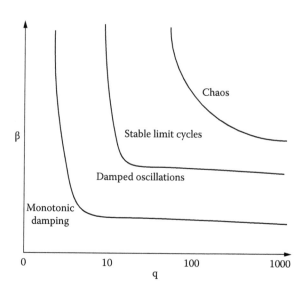

FIGURE 11.13 The dynamic behavior of Equation (11.4). The curves separate the regions of monotonic and oscillatory damping to a stable point, stable limit cycles, and chaos. The thin curve indicates where 2-point cycles give way to higher-order cycles. (Redrawn after Hassel, M.P., et al., *J. Anim. Ecol.*, 45, 471–486, 1976.)

situations? The answer, as seen in Figure 11.15, is that a maximum growth rate of about 0.35–0.40 day^{-1} seems to give favorable conditions for the entire system, as the eco-exergy is at maximum and stable conditions are obtained. A maximum growth rate of more than about 0.65–0.70 day^{-1} gives, however, chaotic situations for the two zooplankton species.

Figure 11.16 shows a similar result for fish when it is included as a state variable; see the conceptual diagram in Figure 11.14. The two zooplankton state variables have been given maximum growth rates of 0.35 and 0.40 day^{-1}. A maximum growth rate of about 0.08–0.1 day^{-1} seems favorable for the fish, but again, too high a maximum growth rate (above 0.13–0.15 day^{-1}) for the state variable fish will give oscillations and chaotic situations with violent fluctuations.

The parameter estimation is often the weakest point for many of our ecological models. The above-mentioned results make it possible to reduce these difficulties by imposing the ecological facts that all the species in an ecosystem have the properties (described by the parameter set) that are best fitted for survival under the prevailing conditions according to ELT. The property of survival can currently be tested by use of eco-exergy, since it is survival translated into thermodynamics. Coevolution, i.e., when the species have adjusted their properties to each other, is considered by application of eco-exergy for the entire system. This method enables us to reduce the feasible parameter range, which can be utilized to facilitate our parameter estimation significantly.

It is interesting that the ranges of growth rate actually found in nature (see, for instance, Jørgensen et al., 1991) are approximately those which give stable, i.e., nonchaotic, conditions. On the other hand, the growth rates found (see Figures 11.13 and 11.14) are slightly less than those that give chaos.

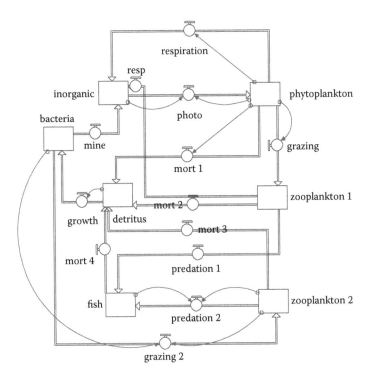

FIGURE 11.14 (See color insert.) Model applied for the parameter examinations that resulted in Figures 11.15 and 11.16.

If we would use lower growth rates, the eco-exergy would be much lower. The resources available—in this case the food for zooplankton—would not be utilized sufficiently. This means that to utilize the resources properly, a high growth rate is needed, and therefore it becomes a growth rate at the edge of chaos.

All in all it seems possible to conclude that the parameters that we can find in nature are usually those that ensure a high probability of survival and growth in all situations; chaotic situations are thereby avoided. The resources are also utilized best by these parameters to obtain the highest possible eco-exergy. The parameters that could give possibilities for chaotic situations have simply been excluded by selection processes. They may give high eco-exergy in some periods, but later the eco-exergy becomes very low due to the violent fluctuations, and it is under such circumstances that the selection process excludes the parameters (properties) that cause the chaotic behavior.

The same results presented above by the use of the model conceptualized in Figure 11.14 can be obtained by even simpler models, consisting, for instance, of phytoplankton, nutrients, detritus, and zooplankton, but dependent on the selection of parameters. It should also be mentioned that the model in Figure 11.14 with the growth rates for the two classes of zooplankton in 0.7 24 h^{-1} is showing deterministic chaos according to Equation (11.2). The model is with these parameter values for the growth

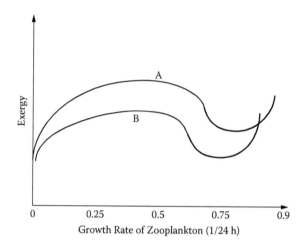

FIGURE 11.15 Eco-exergy is plotted vs. maximum growth rate for the two zooplankton classes in Figure 11.14. A corresponds to the state variable "zoo," and B the state variable "zoo2." The bolded lines after a growth rate of about 0.6 (1/24 h) correspond to chaotic behavior of the model, i.e., violent fluctuations of the state variables and the eco-exergy. The shown values of the eco-exergy above a maximum growth rate of about 0.65–0.7 day⁻¹ are therefore average.

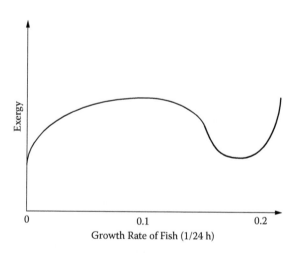

FIGURE 11.16 The exergy is plotted vs. the maximum growth rate of fish. The bolded line after a growth rate of about 0.14 (1/24 h) corresponds to chaotic behavior of the model, i.e., violent fluctuations of the state variables and the exergy. The shown values of the exergy above a maximum growth rate of about 0.13–0.15 day⁻¹ are therefore average values.

rate of the two classes of zooplankton very sensitive to the initial values of the state variables, particularly for the two zooplankton classes.

11.5 Summary of Important Points in Chapter 11

1. Buffer capacity, β, is defined as follows (Jørgensen, 1994a, 2002):

$$\beta = 1/(\partial \text{ (state variable)}/\partial \text{ (forcing function)})$$

   or in the environmental management context:

$$\Delta \text{ forcing function}/\Delta \text{ state variable}$$

2. There is a strong correlation between the sum of many (relevant) buffer capacities and the eco-exergy of the considered ecosystem. Eco-exergy is applied as an indicator for the sum of buffer capacities (resistance) of an ecosystem.

3. Intermediate disturbances favor species diversity and give the ecosystem a high eco-exergy and a high sum of various relevant buffer capacities, while constant conditions give a low diversity and a relatively low eco-exergy and a low sum of buffer capacities.

4. Lakes have two structures: below about 0.06 mg P/L the structure is dominated by zooplankton and carnivorous fish, and above about 0.13 mg P/L the structure is dominated by phytoplankton and planktivorous fish. Between the two indicated concentrations of phosphorus both structures can dominate dependent on the history. This corresponds to a hysteretic behavior, which is important to have in mind in environmental management of lakes.

5. For shallow lakes, below about 0.1 mg P/L submerged vegetation is dominant and above about 0.25 mg P/L phytoplankton is dominant and is able to outcompete the submerged vegetation by shading. Between 0.1 and 0.25 mg P/L both submerged vegetation and phytoplankton can be dominant dependent on the history.

6. The definition of deterministic chaos implies that the distance between the two curves with slightly different initial conditions is growing exponentially:

$$d(t) = d(0)^* e^{L^* t}$$

   where $d(t)$ is the distance at time = t, $d(0)$ is the distance at time = 0, and L is a positive number, called the Lyapunov exponent, which is a quantitative indicator for chaos.

7. Lower growth rates for species than those found in nature or by modeling would yield a lower eco-exergy. The resources available—in this case the food for the species—would not be utilized sufficiently. This means that to utilize the resources properly a sufficiently high growth rate is needed, and it often becomes a growth rate at the edge of chaos.

Exercises/Problems

1. Why is it more difficult to quantify resilience than buffer capacity and resistance?
2. Why is it important to have multidimensional stability concepts?
3. It is shown by modeling and statistics that eco-exergy and the sum of buffer capacities are well correlated. Explain why this correlation is rather obvious.
4. Why would very fast and frequent changes not imply a high species diversity?
5. Are the seasonal changes beneficial for species diversity in the temperate zone?
6. Are the diurnal changes beneficial for the species diversity?
7. How would the lake ecosystem react to the application of biomanipulation if the total P concentration was 0.5 mg/L?
8. A hysteresis behavior of aquatic ecosystems is also possible when nitrogen is the limiting nutrient. In which range would the hysteretic behavior be observed if nitrogen is the limiting nutrient? It is assumed that the N:P ratio for phytoplankton can be used to find the range. See also Chapter 5. The answer should be given for both deeper aquatic ecosystems and shallow aquatic ecosystems with possibilities for submerged vegetation.
9. Consider a model consisting of the food chain nutrients, phytoplankton, zooplankton, and detritus. Draw a conceptual diagram of this model by use of the STELLA symbols. Use a low grazing rate in the firsthand and explain why it is expected that the eco-exergy will increase by increasing the grazing rate. Choose a very high grazing rate and explain why it gives significant changes of the phytoplankton and zooplankton concentrations. Eventually develop the model by use of the STELLA software and make simulations to generate results similar to those in Figures 11.15 and 11.16.

12

The Components of Ecosystems Form Ecological Networks

Life did not take over the globe by combat, but by networking.

—Margulis and Sagan, *Microcosmos*

One of the three growth forms is development of the ecological networks of ecosystems and the efficiencies of the network. The networks make reuse and recycling possible, and it can be shown that they increase the utilization efficiencies of matter and energy. Ecological networks give the ecosystems 13 characteristic properties. Analyses of the networks by the application of matrix calculations are able to quantify the characteristic properties that ecosystems have acquired by the network organization: the mutualism index, the synergism index, the ratio indirect/direct effect, and the homogenization of the effects of the forcing functions, to mention the most important of these properties.

12.1 Introduction

As discussed in Chapter 2, ecosystems recycle about 20 elements that are used to build biomass. Due to the mass conservation principles, all growth and development would stop without the recycling. The recycling is as shown in Figure 2.8 for phosphorus and nitrogen, realized by simple ecological networks. Figures 12.1 and 12.2 give two other examples of larger networks. Figure 12.1 is characteristic for deeper lakes, where a thermocline divides the lake into epilimnion and hypolimnion, and Figure 12.2 is for marine ecosystems. The latter example is modeled by Ecopath. The results of the larger network are formation of more recycling possibilities. The utilization of the energy received from the solar radiation also requires cycling, as illustrated in Figure 3.4. Matter and energy can carry information.

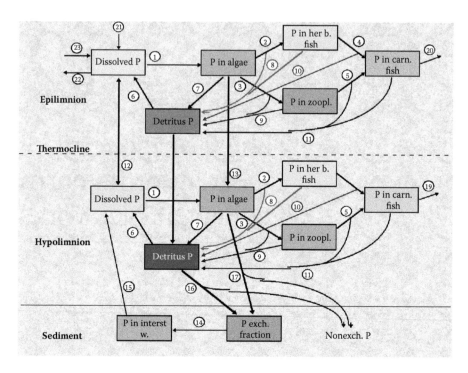

FIGURE 12.1   (See color insert.) Deeper lakes have more comprehensive networks, because the lake is divided by the thermocline into epilimnion and hypolimnion.

Ecological networks therefore become the prerequisite for the indispensable recycling of matter, energy, and information (Patten, 1985). The networks reinforce at the same time the cooperation of the components in the ecological network (Patten, 1978, 1981, 1982, 1991). It is therefore beneficial for the ecosystems to develop ecological networks, and as presented in Chapter 2, ecosystems grow and develop by:

1. Increasing the biomass
2. Increasing the information
3. Development of the ecological networks and their efficiencies

One might say that ecological networks have a synergistic effect in the sense that all the components in the network benefit from the formation of the network (Patten, 1985, 1991, 1992).

In the next section, we will discuss and illustrate by relatively simple examples the advantages of network aggradation. In Section 12.3, we will present 13 cardinal hypotheses about the effect of an ecological network on the ecosystem level. The hypotheses will be presented without going into mathematical details. There will be references to particularly B.C. Patten's work, which has resulted in an important theory for ecological networks, encompassing these 13 cardinal hypotheses and the mathematics applied to analyze the ecological networks. The most important matrix analyses that are used to understand the important properties of ecological networks are presented

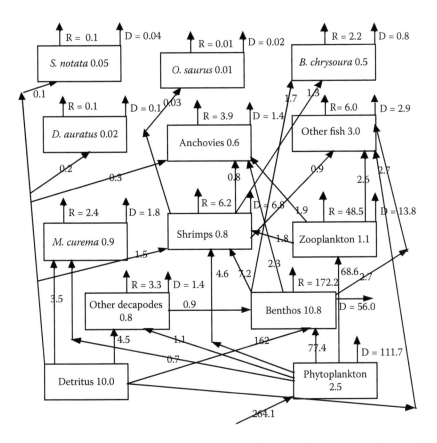

FIGURE 12.2 Example of a marine Ecopath model, taken from Christensen and Pauly (1993). R means respiration and D the transfer to detritus. Notice that all the components are in steady state: input = output. The unit for the state variables is g/m² and for the processes g/(m² year).

in Section 12.4. The most simple analyses that require only a few computations will be shown, while the more complex computations are recommended to be carried out by the downloadable software EcoNet; see http://eco.engr.uga.edu/DOC/econet4.html#scalar. Readers can easily use the software, as it is explained very well on the website. The results of using the software will be presented to give the readers a clear image of all the relevant network properties, which can be determined by EcoNet.

Information about the background of the software can be found in Kazanci (2007, 2009) and Schramski et al. (2011).

The properties of an ecological network are very important for the proper function of ecosystems, including how they can comply with the thermodynamic laws, including ELT.

Ecological networks are a crucial method for ecosystems to move away from thermodynamic equilibrium, and thereby get the most eco-exergy out of the conditions for the ecosystems, including the resources available for the ecosystems to grow and develop.

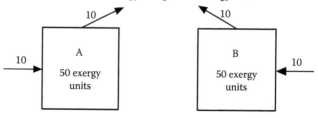

(a) No coupling between A and B. The through-flow is 20
and the exergy storage is 100 exergy units

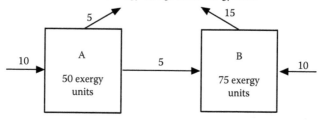

(b) A coupling from A to B. The through-flow is now 25
and the exergy storage is 125 exergy units.

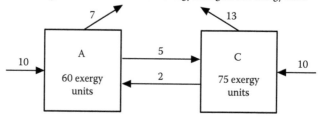

(c) A coupling from A to B and a coupling from B to A. The
through-flow is now 27 and the exergy storage is 135 exergy units

FIGURE 12.3 The two-compartment system illustrates the advantages of network aggradation. The eco-exergy and throughflow increase from no coupling to one coupling to cyclic coupling.

12.2 Ecological Networks Increase Utilization Efficiency of Matter and Energy

The advantages of network aggradation are illustrated in Figure 12.3. Steady state is presumed for this simple illustration, or input = output, and the flows are donor determined by a first-order reaction. The flows and storages could represent the eco-exergy of the simple system. The outflows from the compartments represent the respiration or the energy needed for maintenance of the compartment (the organisms). The retention time is five time units, which implies that the eco-energy is five times the input. As seen, the throughflow and the eco-exergy increase because of the coupling.

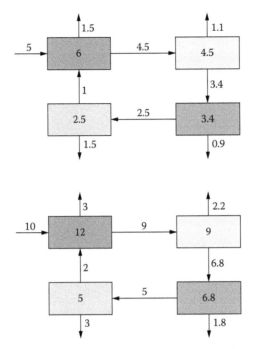

FIGURES 12.4 AND 12.5   (See color insert.) The difference between the two figures is the input of eco-exergy or energy. The networks are at steady state and the flows are first order donor determined. The double input of eco-exergy or energy is able to support twice as much storage in all four compartments.

a: Total eco-exergy is 100 and throughflow is 20. Inputs and outputs are 10 for both A and B. They are usually denoted input environ and output environ to indicate that they are the exchange between the system and its environment.

b: Total eco-exergy is 125 and throughflow is 25

c: Total eco-exergy is 135 and throughflow is 27.

The assumptions, which are the basis for the illustration, are often applied in ecological modeling and give, in many cases, a realistic image of the flows and storages.

Figures 12.4–12.7 show similar results for slightly more complex networks. The examples in Figures 12.3–12.7 show that the formation of networks clearly gives a better utilization. It is presumed in the examples that eco-exergy as input is driving the cycling, but it would also be possible to follow matter or total energy, or even information. From the examples it can be seen that more couplings or more recycling by additional connections in the network increases the utilization efficiency. Network formation offers a great advantage for the energy and matter utilization of ecosystems.

It should, however, be remembered that the network examples are all based on steady state, but ecosystems are often at steady state or close to steady state. First-order reactions, which are also assumed, are also often valid for ecological processes (see, for instance, Jørgensen and Fath, 2011), at least in a narrow range. The Michaelis-Menten equation, which is often used to represent ecological processes, corresponds, for small

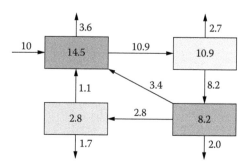

FIGURE 12.6 (See color insert.) An extra cycling from compartment 3 to compartment 1. The result is that the eco-exergy, which is 32.8 in Figure 12.5, increases to 36.4.

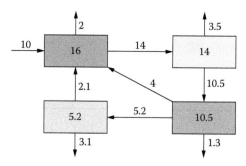

FIGURE 12.7 (See color insert.) The first compartment represents bigger organisms than in Figure 12.6, which implies that the relative loss by respiration is reduced. Therefore, more eco-exergy is available to support the other compartment and recycle. The result is a total eco-exergy of 45.7.

concentrations, to a first-order reaction. The retention time in Figure 12.3 is five time units, which implies that the sum of the compartments is 20% of the total throughflows. The retention time in Figures 12.4–12.7 is one time unit, which means that the sum of the compartments becomes equal to the total throughflows. Anyhow, observations on how ecological networks are working in ecosystems confirm that the networks can explain recycling, reuse, and the relatively high utilization efficiency that is observed in ecosystems (Jørgensen et al., 2007). The network idea was due to the high utilization efficiency being adopted in industries and agriculture (see, for instance, Jørgensen, 2006).

The sum of the compartments is correlated strongly with the throughflow. It is in accordance with ELT: Ecosystems move away from thermodynamic equilibrium and thereby obtain most possible eco-exergy stored in the components and the maximum throughflow (maximum power).

12.3 Cardinal Hypotheses about the Properties of Networks

The 13 cardinal hypotheses are the result of network analyses that will be presented in some detail in the next section. In this section, the 13 hypotheses will be presented together with the ecosystem properties that they can explain.

CH1: Networks allow pathway proliferation. The matter, energy, and information continue to be transported around the system, because the network offers complete recycling. Pathway numbers increase exponentially with increasing pathway length, with the result that the system becomes a complex interconnected network. All components are connected directly or indirectly. In principle, the pathways neither begin nor end. This pathway proliferation is among the sources of an essential holistic understanding of ecosystems.

CH2: Networks imply nonlocality. The amount of matter or energy carried out at any given step is less than that at the previous step due to dissipation—the second law of thermodynamics. The overall rate of decay can be expressed as exponential decay as the radioactivity or biological decomposition. But since the matter or energy continues to cycle and use the pathway proliferation, the total matter or energy transferred between compartments exceeds that of direct intercompartmental transfers. Indirect pathways—those of lengths >2—deliver more matter or energy between compartments than a direct link between them. The influences carried by the indirect transfers are said to be nonlocal.

CH3: The control of networks is distributed. According to the nonlocality presented as CH2, the control of ecosystem processes will have to be dominantly indirect, or expressed differently, the control is distributed, which can be shown to have origins in the openness (Patten and Auble, 1981; Fath 2004).

CH4: Networks homogenize the influences. Due to the nonlocality, there is so much mixing in the entire network that the causalities tend to be evenly spread over the entire interactive network. This means that all compartments are equally significant in generating and receiving influences to and from all the others, or expressed differently, the intercomponent relationships in the network tend to become homogeneous; for further details, see Fath and Patten (1999).

CH5: Networks imply internal amplification. Networks have to follow the second law of thermodynamics, but as has been shown in the examples presented in Section 12.2, the energy and matter utilization efficiencies are increased because of the cycling (Patten, 1985). The otherwise wasted matter and energy are simply reused. This property is clearly rooted in the cyclic interconnection.

CH6: Network unfolding. Network unfolding refers to the proliferation of transfer levels in ecosystems. Higashi et al. (1989) developed a methodology for unfolding a network into corresponding macro chains in contrast to the sequential food chains. The unfolding makes it easier to analyze cycling of energy or matter.

CH7: Networks have synergism. The matter and energy utilization efficiencies are, as illustrated in Section 12.2, increased by networks. Due to the infinite recycling in the network, the wasted matter and energy are minimized. The nonlocal indirect effects exceed the direct effects due to this infinite recycling of energy and matter, as mentioned as hypothesis CH2. This network advantage can be denoted network synergism.

CH8: Network mutualism. The relations of every pair of compartments in the network may be either positive (+), negative (–), or neutral (0), resulting from the direct and indirect transactions that connect them.

The most common types of ecological interactions are:

Predation that is positive for the predator and negative for the prey; it can be indicated as (+,−).
Competition with this terminology would be (−, −).
Mutualism (+, +)
Neutralism (0, 0)
Catabolism or dissipation (−, 0)
Anabolism (+, 0)
Commensalism (0, +)
Amensalism (0, −)

It has been shown by Patten (1991, 1992) that there is a shift to more positive interaction types as a result of the cooperation in a network and the indirect effect. The shift is, of course, dependent on the numerical value of the direct connection/effect. Generally there is a tendency that predation becomes closer to (+, 0), competition closer to (0, 0), neutralism closer to (+, +), catabolism closer to (0, +), anabolism closer to (+, +), commensalism closer to (+, +), and amensalism closer to (0, 0). It is denoted network mutualism. The network synergism and mutualism make nature a beneficial place for life. It is opposite the Darwinism expression "struggle for life," which is true local, but the network approach reveals the global effects and results of the cooperation of the components in a network.

CH9: Network aggradation. The aggradation is already illustrated clearly in Section 12.2. The network results in higher values of eco-exergy for the compartment due to a higher utilization of the matter and the available free energy. The network can, as already stressed, not defeat or violate the second law of thermodynamics, but the amount of free energy lost to heat (respiration) can be reduced by a better utilization of the recycling free energy. Diversity will in this context contribute to more elaborated networks and thereby contribute also to network aggradation. Notice that the network aggradation also contributes to increasing buffer capacities.

CH10: Network boundary amplification. Networks can be developed as discrete trophic levels within the reticular networks. An effective recycling is the result, although the various compartments or nodes of the networks have different properties and process different substances, as is characteristic for primary producers, herbivores, carnivores, top carnivores, and omnivore species. Notice that this hypothesis uses a close analogy between an ecosystem and the metabolic cycle of an organism.

CH11: Networks enfolding. The storages and the indirect energy and matter flows are incorporated into the empirically observed and determined flows and storages. This ability of the network is denoted network enfolding. The calculation of the resulting storages and flows is another challenge, which we will turn to in the next section, while a set of observations at a certain time inevitably will be the result of the continuous recycling in the network, from the time of input.

CH12: Network environ autonomy. The ecosystems are open and have environments that are uniquely attached to them. This implies that not only are the living components in the networks unique, but also the environs (input and output environs) they project. Or expressed differently, both the organisms and their environs are autonomous.

CH13: Networks imply a holistic evolution. This hypothesis claims that both the genomes and the environ code for phenotypes are both seen as heritable elements. The term *envirotype* (see Figure 9.8) has been used to cover this hypothesis. The environ code is considered an ecological inheritance. An example should be mentioned to illustrate this ecological inheritance. It was observed that milk bottles left on the doorsteps of households were opened by birds (blue tits and great tits; see Fisher and Hinde, 1949). The songbirds opened the bottles by tearing off the foil caps and drank the milk from the top of the bottles. This habit was spread by social learning, and this behavior has changed the birds' environment, which of course also changed the selection pressure. Ecological inheritance is the species' ability to change their environment with a consequent change of selection pressure.

To summarize, formations of networks have a significant effect on the properties of ecosystems.

12.4 Network Analyses

The network diagrams in Figures 12.3–12.7 are denoted diagraphs, and to illustrate the most simple network analyses, we will use the diagraph in Figure 12.7. The first applied matrix analysis is the adjacency matrix, which indicates the realized pathways. The realized pathways are marked by 1 in the matrix, and the nonrealized pathways by 0. The columns indicate "from" compartments 1, 2, 3, and 4, and the rows indicate "to" compartments 1, 2, 3, and 4. If a direct arc from j to i exists, then $a_{ij} = 1$. Direct arcs in Figure 12.7 are from 1 to 2, from 2 to 3, from 3 to 1 and 4 and from 4 to 1.

The adjacency matrix for Figure 12.7, A, is shown in Figure 12.8. This matrix allows us to calculate the connectivity, which is the fraction of all possible pathways that are realized. The possible pathways are $4*4 - 4 = 12$, and 5 are realized, which means the connectivity is $5/12 = 0.41$. A connectivity of >0.5 usually means a rigid network, while a connectivity of <0.25 or 0.3 makes a too loose network. A connectivity between 0.3 and 0.5 is preferred. The adjacency matrix indicates the presence of direct connection by 1, and the matrix A^2 gives the number of pathways that take two steps between two compartments. Correspondingly, A^3 gives the number of pathways that take exactly three steps, and so on. A^m therefore gives the number of pathways of the length m.

F is the flow matrix. 1s in A are replaced by the actual flows. The reuse of matter or energy in the network is important for the network properties. The yellow compartment in Figure 12.7 receives 14 units but uses only 3.5 for maintenance, while the remaining 10.5 units are reused in the network. This means that $10.5/14 = 0.75$, or 75% of the matter or energy will be reused and will eventually come back to the yellow compartment.

$$A = \begin{vmatrix} 0 & 0 & 1 & 1 \\ 1 & 0 & 0 & 0 \\ 0 & 1 & 0 & 0 \\ 0 & 0 & 1 & 0 \end{vmatrix}$$

$$F = \begin{vmatrix} 0 & 0 & 4 & 2.1 \\ 14 & 0 & 0 & 0 \\ 0 & 10.5 & 0 & 0 \\ 0 & 0 & 5.2 & 0 \end{vmatrix}$$

$$T = \begin{vmatrix} 16.1 & 14 & 10.5 & 5.2 \end{vmatrix}$$

$$G = \begin{vmatrix} 0 & 0 & 0.38 & 0.31 \\ 0.87 & 0 & 0 & 0 \\ 0 & 0.75 & 0 & 0 \\ 0 & 0 & 0.49 & 0 \end{vmatrix}$$

FIGURE 12.8 The adjacency matrix, A, the flow matrix, F, the throughflow vector, T, and the nondimensional or normalized direct flow matrix, G, for the diagraph (Figure 12.7) are shown.

We distinguish between direct (where the causes and effects are adjacent) and indirect effects. Any transfer of energy or matter propagates and cycles many times in the network, before it finally is dissipated as heat used for maintenance to the environment. It is therefore not surprising that relatively complex calculations are needed to express the role of indirect effects compared with the direct ones. These calculations can be carried out by EcoNet. Let us, however, make some simple calculations to illustrate the indirect effect. If we consider for the yellow compartment in Figure 12.7 that 75% of the matter or energy is reused in the second cycle, then 75% of the 75% or 56.25% will be reused in the third cycle, and 75% of 56.25% in the fourth cycle, or 42.2%, and so on. Notice, however, that we have simplified the effect by not considering other inputs or outputs. This means that

The indirect effect after eight cycles
   = direct effect (0.75 + 0.5625 + 0.422 + 0.3165 + 0.237 + 0.178 + 0.141)
   = 2.6*direct effect   (12.1)

The indirect effects calculated by EcoNet consider an infinite number of cycles and will therefore indicate a higher ratio of indirect to direct effects, but these simple calculations show clearly that indirect effects in many cases will be several times the direct effects. The calculations are more complicated, particularly when we consider many inputs and outputs, but it is here EcoNet can help us to make the calculations considering these complications. The mathematics behind these calculations will not be presented in this textbook, because they are complex and not needed to understand how ecosystems are working and which benefits the ecosystems gain by the network organization.

T gives the throughflow of the four compartments. Notice also that the throughflow is 45.8 and the storage (= the sum of the compartments) is 45.7. The retention time is one unit of time. The difference is due to rounding off errors. With the first-order expression applied for the respiration and the steady state with the retention time one unit, the storage (for instance, eco-exergy stored) and the power (= the total through-flows) should be the same. Correlations between eco-exergy storage and the maximum power (= the sum of the throughflows) are in accordance with ELT: The ecosystem strives toward the highest possible eco-exergy storage under the prevailing conditions, but the highest possible eco-exergy storage implies also the maximum power is reached (see Chapter 7).

The information contained in the F- and G-matrices could be applied as inputs to EcoNet, which is programmed to carry out the complicated calculations needed to reveal all the relevant ecological network properties. The F-matrix gives the direct flows, from compartment 1 in the first column and for compartment 2 in the second column, and so on, and to compartment 1 in the first row, and so on. The arc from j to i is indicated as a_{ij}. The G-matrix indicates the flow relative to the donor storages. This means that, for instance, the flow 14 in Figure 12.7 is in the G-matrix indicated as 14/16, or 0.87; see Figure 12.8.

The software does have extended results, which include control analysis and an eco-exergy calculator. The analysis data can be downloaded in MATLAB® format (.m file) or in spreadsheet format (.csv file). Overall, the EcoNet software provides a very easy to use application for researchers or students interested in conducting network analysis. The results of using EcoNet on the network in Figure 12.7 and with the corresponding matrices are shown in Table 12.1. The software is also able to calculate the eco-exergy, but it would require that the β-values be indicated for the compartment. The software explains further the background knowledge (press the buttons "learn more") about all the results, and particularly the system-wide properties are important to read and understand in this context. The explanation of the system properties is given in Table 12.2.

Notice particularly the following results in Table 12.1:

1. The indirect effects are more than three times the direct ones.
2. The aggradation index is 4.58.
3. The synergism index is 6.73.
4. The mutualism index is 2.2.

In other words, in spite of Figure 12.7 being a simple ecological network, it gives significant advantages to the participants of the ecological network. It is very beneficial for an ecosystem to enhance the growth of the ecological network and increase the efficiencies, expressed by the various indices, of the network. There are, by the growth form 3 (see Section 12.1), enormous possibilities for the ecosystem to move away from the thermodynamic equilibrium.

TABLE 12.1   Results Obtained by the Application of EcoNet in Figure 12.7

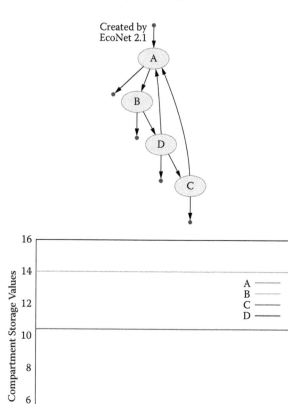

Network Diagram

Data for time-course figure.

Compartmental Properties

	Initial Storage	Final Storage (x)	Environmental Input (z)	Environmental Output (y)
A	16	16	10	2.1
B	14	14	0	3.5
C	5.2	5.2	0	3.1
D	10.5	10.5	0	1.3

TABLE 12.1 (*Continued*) Results Obtained by the Application of EcoNet in Figure 12.7

	Input Throughflow	Output Throughflow	Residence Time
A	16.1	16.1	0.993789
B	14	14	1
C	5.2	5.2	1
D	10.5	10.5	1

System-Wide Properties

Link density	1.25
Connectance	0.3125
Total system throughflow	45.8
Finn's cycling index	0.355569
Indirect effects index	3.03594
Ascendency	58.5963
Development capacity	86.4408
Aggradation index	4.58
Synergism index	6.73482
Mutualism index	2.2
Homogenization index	3.92417

Adjacency Matrix (A)

	A	B	C	D
A	0	0	1	1
B	1	0	0	0
C	0	0	0	1
D	0	1	0	0

Stoichiometric Matrix (SM)

	*→1	1→2	2→4	4→3	4→1	3→1	1→*	2→*	3→*	4→*
A (1)	1	−1	0	0	1	1	−1	0	0	0
B (2)	0	1	−1	0	0	0	0	−1	0	0
C (3)	0	0	0	1	0	−1	0	0	−1	0
D (4)	0	0	1	−1	−1	0	0	0	0	−1

Flow Matrix (F)

	A	B	C	D
A	0	0	2.1	3.99999
B	14	0	0	0
C	0	0	0	5.2
D	0	10.5	0	0

TABLE 12.1 (*Continued*)   Results Obtained by the Application of EcoNet in Figure 12.7

Throughflow Analysis

Normalized Flow Matrix (G)

	A	B	C	D
A	0	0	0.403846	0.380952
B	0.869565	0	0	0
C	0	0	0	0.495238
D	0	0.75	0	0

Throughflow Analysis (N)

	A	B	C	D
A	1.61	0.701499	0.650192	0.935332
B	1.4	1.61	0.565384	0.813332
C	0.52	0.598	1.21	0.797333
D	1.05	1.2075	0.424038	1.61

Normalized Flow Matrix (G′) *output oriented*

	A	B	C	D
A	0	0	0.130435	0.248447
B	1	0	0	0
C	0	0	0	1
D	0	1	0	0

Throughflow Analysis (N′) *output oriented*

	A	B	C	D
A	1.61	0.609999	0.21	0.609999
B	1.61	1.61	0.21	0.609999
C	1.61	1.61	1.21	1.61
D	1.61	1.61	0.21	1.61

Storage Analysis

Partial Turnover Rate Matrix (C)

	A	B	C	D
A	−1.00625	0	0.403846	0.380952
B	0.875	−1	0	0
C	0	0	−1	0.495238
D	0	0.75	0	−1

TABLE 12.1 (*Continued*) Results Obtained by the Application of EcoNet in Figure 12.7

Storage Analysis (S)

	A	B	C	D
A	1.6	0.697142	0.646153	0.929523
B	1.4	1.61	0.565384	0.813332
C	0.52	0.598	1.21	0.797333
D	1.05	1.2075	0.424038	1.61

Partial Turnover Rate Matrix (C′) *output oriented*

	A	B	C	D
A	−1.00625	0	0.13125	0.25
B	1	−1	0	0
C	0	0	−1	1
D	0	1	0	−1

Storage Analysis (S′) *output oriented*

	A	B	C	D
A	1.6	0.609999	0.21	0.609999
B	1.6	1.61	0.21	0.609999
C	1.6	1.61	1.21	1.61
D	1.6	1.61	0.21	1.61

Utility Analysis

Mutual Relations

	A	B	C	D
A	+	−	−	+
B	+	+	+	−
C	−	+	+	+
D	+	+	−	+

Utility Series Matrix (D)

	A	B	C	D
A	0	−0.869565	0.130435	0.248447
B	1	0	0	−0.75
C	−0.403846	0	0	1
D	−0.380952	1	−0.495238	0

TABLE 12.1 *(Continued)* Results Obtained by the Application of EcoNet in Figure 12.7

Utility Analysis (U)

	A	B	C	D
A	0.646781	−0.265404	−0.0627299	0.297014
B	0.469825	0.473169	0.119862	−0.118288
C	−0.0252585	0.455752	0.781877	0.433788
D	0.235941	0.348569	−0.243456	0.553736

	A	B	C	D
A	0	−0.0564286	−0.0596154	−0.0419048
B	0.0564286	0	−0.0746154	−0.0569048
C	0.0596154	0.0746154	0	0.112949
D	0.0419048	0.0569048	−0.112949	0

Control Analysis II (CR)

	A	B	C	D
A	0	0.564286	0.596154	0.419048
B	0	1.11022e-16	0.64883	0.494824
C	0	0	0	0
D	0	0	0.736622	0

TABLE 12.2 Explanation of the Results Obtained by EcoNet Applied in Figure 12.7 (EcoNet is also given the explanations with an indication of relevant references)

System-Wide Properties

Some of these indexes depend on the analysis matrices defined below.

1. **Link density:** Number of intercompartmental links (d) per compartment: d/(number of compartments).
2. **Connectance:** Ratio of the number of actual intercompartmental links (d) to the number of possible intercompartmental links: d/(number of compartments)^2.
3. **Total system throughflow:** Also called total system throughput, it is the sum of throughflows of all compartments: $TST = T_1 + T_1 + \dots + T_n$.
4. **Finn's cycling index:** Measures the amount of cycling in the system by computing the fraction of total system throughflow that is recycled.
5. **Indirect effects index:** Measures the amount of flow that occurs over indirect connections vs. direct connections. When the ratio is greater than 1, indirect flows are greater than direct flows.
6. **Ascendency:** Quantifies both the level of system activity and the degree of organization (constraint) with which the material is being processed in ecosystems.
7. **Aggradation index:** Measures the average path length. In other words, it is the average number of compartments a unit flow quantity (e.g., an N atom, unit biomass, energy quanta, etc.) passes through before exiting the system: $TST/(z_1 + z_2 + \dots + z_n)$.
8. **Synergism index:** Based on utility analysis, it provides a system-wide index for pairwise compartment relations. Values larger than 1 indicate a shift toward positive interactions (mutualism). It is computed as the ratio of the sum of positive entries over the sum of negative entries in the utility analysis matrix U.

TABLE 12.2 (*Continued*)   Explanation of the Results Obtained by EcoNet Applied in Figure 12.7 (EcoNet is also given the explanations with an indication of relevant references)

9. **Mutualism index:** Similar to the synergism index, the mutualism index provides a system-wide index for pairwise compartment relations. Values larger than 1 indicate a shift toward positive interactions (mutualism). It is computed as the ratio of the number of positive entries over the number of negative entries in the mutual relations matrix.
10. **Homogenization index:** Quantifies the action of the network making the flow distribution more uniform. Higher values indicate that resources become well mixed by cycling in the network, giving rise to a more homogeneous distribution of flow.

Adjacency Matrix

Adjacency matrix A consists of 0 and 1 entries and indicates if there exists a direct flow between two stocks. See Figure 12.8 and the explanation in the text for this figure.

Stoichiometric Matrix

Entries of the stoichiometric matrix can be 0, −1, and 1. The entry (i, j) indicates how the storage value of compartment i changes when flow j occurs.

Note that the adjacency matrix contains only intercompartmental relationships. The stoichiometric matrix contains the flows between the system and the environment as well. Therefore, the stoichiometric matrix alone is enough to define network topology.

Flow Matrix

In addition to connectivity information, **F** also provides information on how strong the connections are. As the compartment storage values change in time, the rate of flow between compartments also changes accordingly. Flow rates reach a steady state with the storage values. **F** shows the flow rate between compartments at the end of the simulation. Assuming that the model reaches steady state at the end of the simulation:

If the simulation ends before the system reaches steady state, EcoNet computes the flow matrix based on the final state of the system. Users can use the time-course plot (or throughflow values) to see how close the system is to steady state.

Ecological Network Analysis

The flow matrix **F** represents the connection strength between compartments at steady state. However, even if two compartments are not directly connected, they can still affect each other through indirect connections involving other compartments. For example, there are no direct flows between compartments A and D in the next figure below. However, any change in compartment A will eventually affect compartment D. In this regard, even the model diagram below can be viewed as misleading because it only represents the direct flows. Particularly in well-connected systems, there is often a greater contribution from indirect flows than from direct. This property is called the *dominance of indirect effects*.

EcoNet uses network environ analysis (NEA) to quantify the actual relations among compartments, environmental inputs, and outputs. Unlike most analysis methods, NEA treats the system as a whole and provides an elegant way to quantify the effects of all indirect flows in the system. NEA is not a one-step analysis, but a series of algebraic operations resulting in scalar and matrix values representing various system-wide properties. The first version of EcoNet performed *storage (S)*, *throughflow (N)*, and *utility (U)* analyses. EcoNet 2.0 added many new analysis results.

Some of the analysis results are valid only when the system is close to steady state. EcoNet displays results based on the final state of the system. To get accurate results, users should use the time-course plot (or throughflow values) to make sure that the system is close to steady state by the end of the simulation.

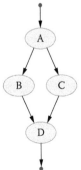

Throughflow Analysis

The definition of throughflow is given compartmental properties. N_{ij} represents how much of the environmental input to compartment j is received by compartment i.

TABLE 12.2 (Continued)   Explanation of the Results Obtained by EcoNet Applied in
Figure 12.7 (EcoNet is also given the explanations with an indication of relevant references)

Note that this input may reach compartment i through indirect flows that involve many other compartments, and N_{ij} accounts for all such possible paths.

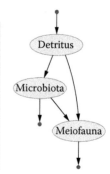

Throughflow Analysis (N: z –> T)			
–	Detritus	Microbiota	Meiofauna
Detritus	1	0	0
Microbiota	0.428571	1	0
Meiofauna	0.991597	0.980392	1

Let's consider the first column of matrix **N** given in the figure above. Obviously 100% of the environmental input to detritus is received by detritus; therefore $N_{11} = 1$. Coefficient $N_{21} = 0.428$ means that 42.8% of the environmental input to detritus is received by microbiota, then 57.2% flows to meiofauna. Coefficient $N_{31} = 0.991$ represents the fact that meiofauna eventually receives 99.1% of the environmental input (to detritus), through microbiota and directly from detritus. Then we can also conclude that 0.9% of the environmental input returns back to the environment through dissipation at microbiota.

To compare the effect of direct and indirect paths from an input to a compartment, EcoNet displays another matrix **G**, defined as follows:

$$G_{ij} = F_{ij}/T_j$$

G can be viewed as a normalized **F** matrix, where each entry F_{ij} is divided by the diagonal element of the row to which it belongs. While F_{ij} represents the actual flow rate from j to i at steady state, G_{ij} represents the ratio of the throughflow of j received by i. Note that both **N** and **G** contain normalized values representing how energy (or biomass, nutrients, C, N, P, etc.) is distributed among compartments. However, **N** accounts for all possible direct and indirect flows among compartments, while **G**, by definition, only represents direct flows among compartments.

Comparing **N** and **G** matrices (previous two figures), we see that entries of both matrices match except for:

$$0.991 = N_{31} = G_{31} = 0.571$$

This is expected for a simple three-compartment model where there is one indirect flow. There are two paths from compartment 1 (detritus) to compartment 3 (meiofauna):

Direct flow	Detritus → meiofauna
Indirect flow	Detritus → microbiota → meiofauna

Both are accounted for in $N_{31} = 0.991$, while $G_{31} = 0.571$ only represents the direct flow. Then we can conclude that $100 \times (0.991 – 0.571)/0.991 = 42\%$ of the interaction between detritus and meiofauna occurs over the indirect connection.

EcoNet provides two more matrices in its "Extended Results": G′ and N′, titled as "input oriented." Similar to G, G′ is defined as follows:

$$G'_{ij} = F_{ij}/T_i$$

TABLE 12.2 *(Continued)* Explanation of the Results Obtained by EcoNet Applied in Figure 12.7 (EcoNet is also given the explanations with an indication of relevant references)

G′ can be viewed as a normalized **F** matrix; where each entry F_{ij} is divided by the diagonal element of the column to which it belongs. N'_{ij} represents the amount of throughflow generated at compartment j for a unit output from compartment i. **N** and **N** are computed as follows:

$$N = (I - G)^{-1}$$

$$N' = (I - G')^{-1}$$

Storage Analysis

Storage analysis matrix S represents the relation between input flow rates and compartment storage values. In the figure above, only detritus receives a direct environmental input. This input is partially transferred to microbiota and meiofauna through indirect flows. S_{ij} represents how much of the storage value of compartment i is contributed by a unit of direct environmental input to compartment j.

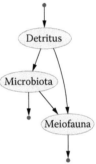

Storage Analysis (S: z –> x)			
–	Detritus	Microbiota	Meiofauna
Detritus	2.8574	0	0
Microbiota	0.840336	1.96078	0
Meiofauna	4.31129	4.26257	4.34783

Let's consider the first row of the storage matrix S:

$$S_{11} = 2.857, S_{12} = S_{13} = 0$$

There are no environmental inputs into microbiota and meiofauna; therefore the storage value of detritus depends only on its own environmental input ($S_{12} = S_{13} = 0$). The environmental input rate to detritus is *10 units/time*, and the steady-state storage value of detritus is *28.57 units*. Therefore, *1 unit* environmental input to detritus contributes to $S_{11} = 2.857$ units of storage value in detritus.

Similarly, the storage value of microbiota (*8.4 units*) is all contributed by the environmental input to detritus; therefore $S_{21} = 8.4/10 = 0.84$. Although the environmental input to microbiota is *0 units/time* in our model, $S_{22} = 1.96$ means that each unit of direct environmental input to microbiota would contribute to $S_{22} = 1.96$ units of storage value. The fact that $S_{23} = 0$ informs us that an environmental input to meiofauna will never affect the storage value of microbiota.

We should note that if there were an additional flow from meiofauna to detritus, there would be no zero entries in **S**, as any environmental input would cycle through all compartments, contributing to all storage values.

EcoNet provides "input oriented" storage analysis matrix S′ in "extended results." S'_{ij} represents how much of the storage has to be generated at compartment j for a unit output from compartment i. **S** and **S**′are computed as follows:

TABLE 12.2 *(Continued)* Explanation of the Results Obtained by EcoNet Applied in Figure 12.7 (EcoNet is also given the explanations with an indication of relevant references)

$$S = -C^{-1}$$

$$S' = -C'^{-1}$$

where

$$C_{ij} = F_{ij}/x_j$$

$$C_{ii} = T_i/x_i$$

and

$$C'_{ij} = F_{ij}/x_i$$

$$C_{ii} = T_i/x_i$$

Utility Analysis

Utility analysis quantifies "mututal relations" among compartments, again including indirect effects. Here's a simple example:

$$* \rightarrow tree \rightarrow deer \rightarrow wolf \rightarrow *$$

Here, * represents the environment. *Deer* and *wolf* have a (–, +) relationship, same as the *tree-deer* relationship. Although not connected with a direct flow, the *tree* and the *wolf* have a mutualistic (+, +) relationship. Figuring out such relations is straightforward for simple models, but extremely difficult when models involve feedback loops, cycling, or cross-level feeding. Utility analysis provides this relation among all compartments regardless of model complexity.

Utility Analysis (U)			
–	Detritus	Microbiota	Meiofauna
Detritus	0.229593	–0.470469	0.128122
Microbiota	–0.297411	0.616033	–0.220334
Meiofauna	0.270242	–0.43321	0.157403

Utility Relations (sign of U)			
–	Detritus	Microbiota	Meiofauna
Detritus	+	–	+
Microbiota	–	+	–
Meiofauna	+	–	+

TABLE 12.2 *(Continued)* Explanation of the Results Obtained by EcoNet Applied in Figure 12.7 (EcoNet is also given the explanations with an indication of relevant references)

The second matrix above represents the relations among compartments in +/– format, while the first matrix **U** provides the quantitative relation strength as well. The three-compartment model we provide here is too simple to present the strength of utility analysis. We postpone this discussion to the next section, where we analyze John M. Teal's Georgia salt marsh model. **S** and **S′** are computed as follows:

$$U = (I - D)^{-1}$$

where

$$D_{ij} = (F_{ij} - F_{ji})/T_i$$

Eco-Exergy Calculator

Eco-exergy is a thermodynamic indicator for ecosystems. It measures a system's deviation from chemical equilibrium. The definition of eco-exergy involves Kullback's information measure. Therefore one needs the "β-values" for the compartments used in their model. Once the user enters these values in the provided spaces, eco-exergy will be computed instantaneously and displayed on the last line. A table for β-values for common compartments and more information on eco-exergy is available in the references.

Example 12.1

Below is shown the network of an intertidal oyster reef taken from Patten (1985). The example has been used many times to illustrate the importance of networks for ecosystems; see, for instance, Jørgensen (2002) and Jørgensen et al. (2007).

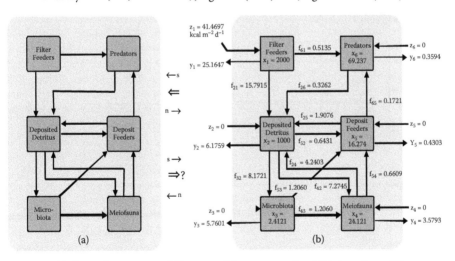

(a) shows the conceptual diagram of the oyster reef and (b) the values of the storages (kcal/m²) and flows (kcal/m²d). s and n indicate that it is both the sufficient and necessary information that can be found on (b).

Use the EcoNet to find:

1. The nondimensional direct flow matrix (G)
2. The integral flow matrix
3. The partial turnover rate matrix
4. The integral storage matrix
5. The mutualism relations
6. The utility matrix (denoted U)

Determine furthermore the connectivity, total throughflow, total storage, indirect effects index, aggradation index, synergism index, and mutualism index, which all can be determined by EcoNet.

Interpretation of the results from a network theoretical point is also required.

Solution

Nondimensional, direct flow matrix:

$$
G = \begin{bmatrix}
0 & 0 & 0 & 0 & 0 & 0 \\
0.38 & 0 & 0 & 0.5 & 0.76 & 0.475 \\
0 & 0.367 & 0 & 0 & 0 & 0 \\
0 & 0.327 & 0.148 & 0 & 0 & 0 \\
0 & 0.027 & 0.148 & 0.0779 & 0 & 0 \\
0.014 & 0 & 0 & 0 & 0.0687 & 0
\end{bmatrix}
$$

Integral flow matrix:

$$
N = \begin{bmatrix}
1.000 & 0 & 0 & 0 & 0 & 0 \\
0.536 & 1.386 & 0.277 & 0.779 & 1.01 & 0.658 \\
0.197 & 0.510 & 1.102 & 0.286 & 0.404 & 0.242 \\
0.205 & 0.529 & 0.253 & 1.297 & 0.419 & 0.251 \\
0.059 & 0.154 & 0.190 & 0.164 & 1.122 & 0.073 \\
0.0185 & 0.0106 & 0.013 & 0.0113 & 0.077 & 1.005
\end{bmatrix}
$$

Partial turnover rate matrix:

$$
C = \begin{bmatrix}
-0.0208 & 0 & 0 & 0 & 0 & 0 \\
0.0079 & -0.0223 & 0 & 0.1758 & 0.1172 & 0.0047 \\
0 & 0.0082 & -3.388 & 0 & 0 & 0 \\
0 & 0.0073 & 0.5 & -0.3516 & 0 & 0 \\
0 & 0.0006 & 0.5 & 0.0274 & -0.1542 & 0 \\
0.0003 & 0 & 0 & 0 & 0.0106 & -0.0099
\end{bmatrix}
$$

Integral storage matrix:

$$S = \begin{bmatrix} 48.0769 & 0 & 0 & 0 & 0 & 0 \\ 24.0392 & 62.172 & 12.423 & 34.929 & 49.283 & 29.516 \\ 0.0582 & 0.150 & 0.325 & 0.0845 & 0.119 & 0.071 \\ 0.582 & 1.505 & 0.721 & 3.690 & 1.193 & 0.714 \\ 0.386 & 0.997 & 1.231 & 1.066 & 7.276 & 0.473 \\ 1.870 & 1.068 & 1.318 & 1.141 & 7.790 & 101.517 \end{bmatrix}$$

Mutualism relations:

$$\text{sgn}(U) = \begin{bmatrix} + & - & + & + & - & - \\ + & + & - & - & + & - \\ + & + & + & - & - & + \\ + & + & + & + & - & + \\ - & + & + & + & + & - \\ + & - & + & + & + & + \end{bmatrix}$$

Utility analysis:

$$U = \begin{bmatrix} 0.832 & -0.221 & 0.0699 & 0.0126 & -0.027 & -0.014 \\ 0.424 & 0.599 & -0.193 & -0.0357 & 0.065 & -0.0012 \\ 0.394 & 0.547 & 0.741 & -0.200 & -0.061 & 0.0071 \\ 0.208 & 0.287 & 0.0014 & 0.946 & -0.056 & 0.0054 \\ -0.0035 & 0.065 & 0.446 & 0.169 & 0.911 & -0.0615 \\ 0.473 & -0.449 & 0.251 & 0.066 & 0.156 & 0.975 \end{bmatrix}$$

EcoNet has given the following results:

Connectivity: 0.3333
Total system throughflow (TST): 83.60
Total storage: 3,112
Indirect effects index: 1.530
Aggradation index: 83.6/41.47 = 2.02
Synergisms index: 6.538
Mutualism index: 2

Interpretation of the Results

Connectivity as it should be between 0.3 and 0.5.
The indirect effects are 53% higher than the direct ones.
The network clearly gives the ecosystems mutualism and synergism.

12.5 Network Selection by Ecosystems

The ELT entails that as the species and the combination of species that gives the ecosystem the highest eco-exergy will be selected, also the network that would give the ecosystem the highest eco-exergy will be selected. The network with the highest eco-exergy of the components will also give the highest flows and total system throughflows, and thereby give the network the best effect of mutualism, synergism, reuse, recycling, aggradation, and the indirect/direct effect ratio. The SDM was introduced in Chapter 7 to illustrate that it is feasible to construct models that are able to change the parameters of the model—this means the properties of the biological components—in accordance with the selection of the best survivors. Models that select the networks that are able to move the systems most away from thermodynamic equilibrium have not yet been developed. If we need to predict more drastic changes of ecosystems, including changes of entire networks in the future due to, for instance, significant climatic changes, it will be necessary to develop what we could call network dynamic models (NDMs). Such models should account for selection of new networks caused by significant changes of the forcing functions, in addition to changes of the species' properties due to shifts in species' compositions and adaptations.

It is, however, interesting to ask the relevant question: Which network changes are actually increasing the eco-exergy? Jørgensen and Fath (2006) have examined some aspects of this question in addition to what we have already presented in Section 12.2. The applied network in this context is shown in Figure 12.9. The dotted lines indicate flows that are added to record the effect of these extra pathways: from carnivores in chain B to top carnivores in chain A, from herbivores in chain A to carnivores in chain B, from detritus to carnivores in chain A, and an extra loop from detritus to herbivores in chain B. All the versions of the network have been modeled by STELLA and run to steady state. Removal of top carnivores and of links from detritus have also been tested. The eco-exergy at steady state was recorded for each of the models. The results are shown in Table 12.3, where the 10 applied changes are listed.

The results can be summarized in the following general propositions, based on interpretation of the network changes listed in Table 12.3 and supplemented by changes of some of the rates:

1. Increased input gives a proportional increase of eco-exergy and power.
2. Addition of extra links will increase reuse and recycling and imply a higher eco-exergy and power (TST) of the network.
3. Additional links will, on the other hand, only affect the power and eco-exergy when it gives additional exergy or energy transfer. Removal of one energy flow from one food chain to another has no effect.
4. Prolongation of the food chain has a positive effect on the power (TST, total system throughflow) and eco-exergy of the network.
5. Reduction of loss of eco-exergy to the environment or as detritus (with a β-value of 1 only) yields a higher power and eco-exergy of the network.

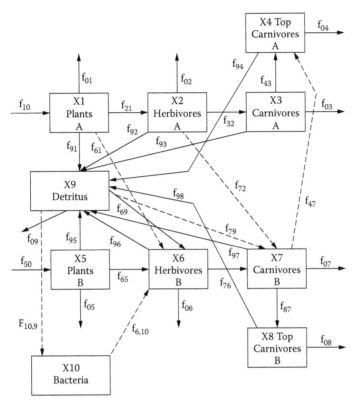

FIGURE 12.9   The network applied to assess the results of network changes (see Table 12.3). Two food chains are included, denoted chain A and chain B. Detritus is jointly for the two chains. The dotted lines indicate flows that are added to record the effect of these extra pathways: from carnivores in chain B to top carnivores in chain A, from herbivores in chain A to carnivores in chain B, from detritus to carnivores in chain A and an extra loop from detritus to herbivores in chain B.

TABLE 12.3   Network Changes and the Corresponding Eco-Exergy Changes at Steady State

Change	Eco-Exergy in Detritus Eqv.	Throughflow
1. Basic model	54,463	58.4
2. Doubling of input exergy to 2	105,636	90.9
3. Removing top carnivores	52,818	45.4
4. Link plants A to herbivores B (specific rate 0.25)	57,424	59.9
5. Link herbivores A to carnivores B	55,567	59.0
6. Link carnivores B to top carnivores A	54,652	58.7
7. Transfer of exergy from A to B	54,463	58.4
8. Adding bacterial food chain	61,938	65.3
9. Detritus to carnivores B instead of to herbivores B	54,762	58.9
10. No link from detritus to herbivores or carnivores B	36,416	50.1

6. Faster cycling—by either faster detritus decompositions or increased transfer rates between two trophic levels—yields higher power and eco-exergy.
7. Input of additional eco-exergy or energy cycling flows has a greater effect the earlier in the food chain the addition takes place.

These rules about the effect of network changes on the eco-exergy and power of the ecosystem are useful when two or more networks are compared, for instance, in a modeling context. The inclusion of top carnivores and the bacterial food chain in models or by assessment of ecosystem health or ecosystem services may be important, as shown by the referred network examination. Consequently, it is recommended before a final model is selected to consider one or more network changes.

12.6 Summary of Important Points in Chapter 12

1. Ecological networks are crucial tools for ecosystems to move away from thermodynamic equilibrium and thereby get the most eco-exergy out of the conditions for the ecosystems, including the resources available for the ecosystems to grow and develop.
2. More couplings or more recycling by additional connections in the network increases the utilization efficiency. A network's formation offers a great advantage for the energy and matter utilization of ecosystems.
3. Networks imply an infinite recycling and that network control is nonlocal, distributed, and homogenized.
4. The effect of networks on ecosystems is significant: synergism, mutualism, amplifications of boundaries, and aggradation (TST > inflows). The effects can be found by EcoNet.
5. Prolongation of the food chain has a positive effect on the power (TST, total system throughflow) and eco-exergy of the network.
6. Reduction of loss of eco-exergy to the environment or as detritus (with a β-value of 1 only) yields a higher power and eco-exergy of the network.
7. Faster cycling—through either faster detritus decomposed or increased transfer rates between two tropic levels—yields higher power and eco-exergy.
8. Input of additional eco-exergy or energy cycling flows has a greater effect the earlier in the food chain the addition takes place.

Exercises/Problems

1. Analyze by EcoNet the network in Figures 12.5 and 12.6 and comment on the results.
2. Mention in addition to ecosystems other systems that use the advantages of networks.
3. Explain how industrial systems could benefit from networks.
4. Explain how agricultural systems could benefit from networks.
5. Develop an adjacency matrix for Figure 12.9, considering only the links indicated with full lines. Calculate the connectivity and comment on the result.

6. Explain under which circumstances TST and eco-exergy are closely correlated. Explain why TST or power generally under all circumstances should be more or less correlated?

7. Explain why networks that have larger biological components result in higher eco-exergy and power.

13

Ecosystems Have a Very High Content of Information

Life is information.

Ecosystems have an enormous amount of information embodied in the genomes of the individuals and in the networks. The concept of ascendency can be used to express the amount of information embodied in the networks, but to comply with the information in the genomes it is necessary to express the flows by using eco-exergy. The evolution can be described as an increase in the information. The increase in information in the genomes is considered the vertical evolution, and the increase of biodiversity, and thereby of the networks and the information that they are carrying, is denoted the horizontal evolution. While the growth of the biomass is close to its limit, the growth possibilities of genetic information and of networks are very far from its limits. Information is embodied in life processes and life is information. Information is not conserved. Transfer of information is irreversible and exchange of information is communication.

13.1 The Information Embodied in the Genes

As presented in Chapter 7, the eco-exergy density can be found from the concentrations of the ecosystem components and their β-values, which express the information content (see Table 4.1):

$$\text{Eco-exergy} = \sum_{i=1}^{i=n} \beta_i c_i \text{ as detritus equivalents at the temperature } T = 300 \text{ K}$$

Specific exergy is exergy relative to the biomass and for the i-th component: $\text{Sp.-ex.}_i = \text{Ex}_i / c_i$. This implies that the total specific exergy per unit of area or per unit of volume of the ecosystem is equal to RTK, where K is the Kullbach's measure of information (see

241

Chapter 3), or expressed by the number of amino acids coded by the genes, AMS, equal to $4.00*10^{-6}*$AMS.

As the average β-value for many ecosystems typically is between 50 and 250, the amount of information carried by ecosystems corresponds to 50–250 times as much as the energy of the biomass. The biomass per m² is for most ecosystems between 1 and 1,000 kg, which would correspond to about 18.7 MJ and 18.7 GJ per m². This implies that the information content will correspond to on the order of between 1.87 and 1,870 GJ per m² if we use an average information content of 100. This means that the number of amino acids coded in an ecosystem is enormous, but it is the information that is controlling all the life processes of the ecosystem.

An ecosystem is a typical middle number system, which implies that most ecosystems will have between 10^{15} and 10^{20} individuals with different properties, different genomes, and different reactions to the prevailing conditions of the ecosystem. The individuals may be presented by 1,000 to 100,000 species, which implies that the ecosystem would have 1,000 to 100,000 completely different genomes. The species and the individuals embody another set of information, which is crucial for the ecosystem and its reaction to the enormous variability of the forcing functions, determining the prevailing conditions.

The evolution has increased the amount of information, as has been shown in Chapter 7. Both the biodiversity and the genomes have increased, when we exclude the last few hundred years, where human beings have reduced the area of nature and the biodiversity.

13.2 The Ascendency

The ascendency represents the information stored in the network, and the definition is shown Figure 13.1.

Example 13.1

Figure 13.2 is used to illustrate the calculations of ascendency. The example is presented by Ulanowitz (see Jørgensen et al., 2007). Calculate the ascendency of the network shown in Figure 13.2 according to definition given in Figure 13.1.

In more quantitative terms, the formula for ascendency is

$$A = \sum_{i,j} T_{ij} \log\left(\frac{T_{ij} T_{..}}{T_{i.} T_{.j}}\right)$$

T_{ij} is the flow from compartment i to compartment j. T_i is compartment i and T_j is compartment j. T is the sum of all flows.

FIGURE 13.1   Definition of ascendency.

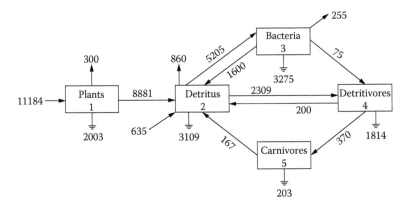

FIGURE 13.2 The ecological network used to illustrate the calculations of ascendency according to the definition in Figure 13.1. The energy exchanges (kcal m^{-2} year^{-1}) among the five major compartments of the Cone Spring ecosystem are indicated.

Solution

The TST of Cone Spring is simply the sum of all the arrows appearing in the diagram. Systematically, this is calculated as follows:

$$\text{TST} = T_{01} + T_{02} + T_{12} + T_{16} + T_{17} + T_{23} + T_{24} + T_{26} + T_{27} + T_{32} + T_{34} + T_{36} + T_{37} + T_{42} + T_{45} + T_{47} + T_{52} + T_{57}$$

$$= 11{,}184 + 635 + 8{,}881 + 300 + 2{,}003 + 5{,}205 + 2{,}309 + 860 + 3{,}109 + 1{,}600 + 75 + 255 + 3{,}275 + 200 + 370 + 1{,}814 + 167 + 203 = 42{,}445 \text{ kcal m}^{-2} \text{ year}^{-1}$$

The subscript 0 represents the external environment as a source, 6 denotes the external environment as a receiver of useful exports, and 7 signifies the external environment as a sink for dissipation.

$$A = \sum_{i,j} T_{ij} \log\left(\frac{T_{ij}T}{\sum_{k}T_{kj}\sum_{q}T_{iq}}\right)$$

Referring again to the Cone Spring ecosystem network shown above, we notice that each flow in the diagram generates exactly one term in the indicated sums. Hence, we see that the ascendency consists of the 18 terms:

$$A = T_{01} \log\left(\frac{T_{01}T}{\sum_{k}T_{k1}\sum_{q}T_{0q}}\right) + T_{02} \log\left(\frac{T_{02}T}{\sum_{k}T_{k2}\sum_{q}T_{0q}}\right) + \ldots + T_{57} \log\left(\frac{T_{57}T}{\sum_{k}T_{k7}\sum_{q}T_{5q}}\right)$$

$$= 20{,}629 - 1481 + 13{,}796 - 94 - 907 + 9{,}817 + 4{,}249 + 1{,}004 + 446 + 295 - 147 + 142 + 4{,}454 - 338 + 1{,}537 + 2{,}965 + 123 + 236$$

$$= 56{,}725 \text{ kcal-bits m}^{-2} \text{ year}^{-1}$$

The calculations of ascendency are sometimes supplemented by calculations of the diversity of flows, the system development capacity, and the system's overhead. The equations to calculate these supplementary indices are given below: While ascendency measures the degree to which the system possesses inherent constraints, we also wish to have a measure of the degree of flexibility that remains in the system. To assess the degrees of freedom, we first define a measure of the full diversity of flows in the system. To calculate the full diversity, we apply the Boltzmann formula to the joint probability of flow from i to j, T_{ij}/T, and calculate the average value of that logarithm. The result is the familiar Shannon formula:

$$H = -\sum_{i,j}\left(\frac{T_{ij}}{T}\right)\log\left(\frac{T_{ij}}{T}\right) \tag{13.1}$$

where H is the diversity of flows. Scaling H in the same way we scaled A, i.e., multiplying H by T, yields the system development capacity, C, as

$$C = -\sum_{i,j} T_{ij}\log\left(\frac{T_{ij}}{T}\right) \tag{13.2}$$

Now, it can readily be proved that $C \geq A \geq 0$, so that the residual $(C - A) \geq 0$ as well. Subtracting A from C and algebraically reducing the result yields the residual ϕ, which we call the systems overhead, as

$$\Phi = C - A = -\sum_{i,j} T_{ij}\log\left(\frac{T_{ij}^2}{\sum_k T_{kj}\sum_q T_{iq}}\right) \tag{13.3}$$

The overhead gauges the degree of flexibility remaining in the system (Ulanowicz, 1986).

Just as we substituted the values of the Cone Spring flows into the equation for ascendency, we may similarly substitute into this equation for overhead to yield a value of 79,139 kcal-bits m^{-2} year^{-1}. Similarly, substitution into the formula for C yields a value of 135,864 kcal-bits m^{-2} year^{-1}, demonstrating that the ascendency and the overhead sum exactly.

It is presumed that the networks in Figures 12.4–12.7 all represent energy storages (for instance, in kJ) and energy flows (for instance, in kJ/24 h); the ascendency of these four networks is not adding much information compared with the information embodied in the organisms (see the values of the information embodied in the genes in Section 13.1). The storages and flows are, however, eco-exergy, and it would therefore be natural to change the values accordingly; see Jørgensen and Ulanowitz (2009).

Let us presume that all the networks in Figures 12.4–12.7 represent an aquatic ecosystem, where the first component is phytoplankton, meaning $\beta = 20$; the second components is zooplankton, $\beta = 100$; the third component is fish, $\beta = 500$; and the fourth component is detritus, $\beta = 1.0$. Figure 12.5 will thereby be changed to Figure 13.3, Figure 12.6 to Figure 13.4, and Figure 12.7 to Figure 13.5.

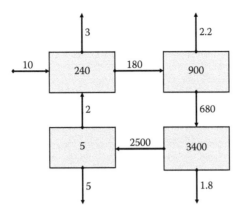

FIGURE 13.3 (See color insert.) Same network as in Figure 12.5, but eco-exergy is applied instead of energy.

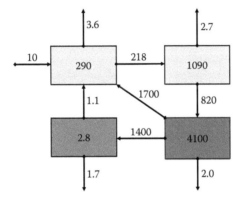

FIGURE 13.4 (See color insert.) Same network as in Figure 12.6, but eco-exergy is applied instead of energy.

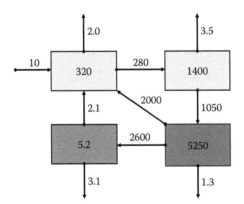

FIGURE 13.5 (See color insert.) Same network as in Figure 12.7, but eco-exergy has replaced energy.

TABLE 13.1 Results of Calculations

Figure Eco-Power[a]	Energy or Eco-Exergy Storage	Energy or Eco-Exergy Ascendency[a]	Energy or Exergy
12.4	16.4	34.65	16.4
12.5	32.8	69.29	32.8
12.6	36.4	65.10	36.4
12.7	45.7	86.92	45.8
13.3	4,545	3,553	3,372
13.4	5,483	4,168	4,149
13.5	6,975	4,666	5,942

[a] If the eco-exergy unit is kJ/m^2, the unit for ascendency and power could be kJ/24 h m^2.

TABLE 13.2 Increases in Energy or Eco-Exergy Storage, Ascendency, and Power

From Figure to Figure	Storage	Ascendency	Power
12.5 to 12.6	11.0%	−6.0%	11.0
12.6 to 12.7	25.5%	33.5%	25.5
13.3 to 13.4	20.6%	17.3%	23.0%
13.4 to 13.5	27.2%	12.0%	43.2%
12.5 to 12.7	39.6%	27.1%	39.6%
13.3 to 13.5	53.4%	31.3%	76.2%

The energy or eco-exergy storage, the energy and eco-exergy ascendency, and the energy and eco-exergy power have been calculated based upon the Figures 12.4–12.7 and 13.3–13.5. The results are shown in Table 13.1. See also Jørgensen and Ulanowitz (2009).

The results show, not surprisingly, that it is necessary to calculate the eco-exergy-based ascendency, if it has to be compared with the eco-exergy storage.

The eco-exergy ascendency should therefore be added to the eco-exergy storage to express what is obtained by changing the network. The sum could be applied as a goal function in structurally dynamic modeling to account for the network changes in addition to changes in species composition.

The increases in percentage for different changes have been calculated (see Table 13.2). The increase from Figure 12.4 to 12.5 is not included in the table because it is just a doubling of all flows, exergy storage, ascendency, and power.

The eco-exergy storage increases significantly more than the energy storage, 20.6% vs. 11.0%, by adding an extra transfer from component 3 to 1. The reason is the increased cycling is utilized particularly by the later components in the food chain, and they have a higher β-value. This is in accordance with the network rules published by Jørgensen and Fath (2006) and presented in Section 12.5. The eco-exergy storage increases 27.2% by adding more information, and thereby reduces the respiration from Figure 13.4 to

Figure 13.5. It is slightly more than for energy; see Figures 12.6 and 12.7. More energy or eco-exergy is available for the flows by reduced respiration loss, which is able to explain the power increasing by this change of the network (Figures 12.6 and 12.7 and Figures 13.4 and 13.5) more (namely, 76.2%) than the energy or the eco-exergy stored.

There is a drop in ascendancy from the network in Figure 12.5 (69.29) to that in Figure 12.6 (65.13). The reason for the drop is that the new flow from 3 to 1 adds ambiguity to the network. Notice that ascendancy is not simply proportional to the total system throughput. If all possible flows are realized in the network, the ascendancy is zero, because ambiguity about where a quantum in any compartment will flow next is maximal. It doesn't matter how strong the flows are, the ascendancy will remain zero, so long as all the magnitudes remain equal.

The very idea behind ascendancy was to modify the total system throughput to quantify how organized (definitive) flow in the system is. The three networks in Figures 12.5–12.7 all have the same total input (10), but their ascendencies are different and reflect the degree of unambiguity in each one.

Now, looking back at Figures 12.5 and 12.6, we notice that flow in Figure 12.5 is not very ambiguous. In particular, if one is in compartment 3, the only other compartment to which quanta can flow is 4. (Of course, it could also leave the system as an export.) In Figure 12.6, by contrast, if a quantum is in compartment 3, there is some uncertainty as to whether it will flow to 4 or 1. This lowers the ascendancy, even though there is more total flow in Figure 12.6 than in Figure 12.5.

Notice how each flow generates only one term in the formula for ascendancy. In particular, the contribution of T_{31} in Figure 12.6 to the ascendancy is

$$A_{31} = T_{31} \log\left(\frac{T_{31}T_{..}}{T_{3.}T_{.1}}\right) \tag{13.4}$$

or

$$A_{31} = 3.4 \log\left(\frac{3.4 \cdot 46.4}{8.2 \cdot 14.5}\right) = 3.4 \log(1.327) = 1.388 \tag{13.5}$$

That is, T_{31} contributes proportionately less than its magnitude to the ascendancy.

T_{34} (= 2.8) contributes 7.001 to the ascendancy. So the total of T_{31} and T_{34} is 8.389. Contrast this to the amount that T_{34} in Figure 12.5 contributes (= 13.27), and the shortfall in Figure 12.6 becomes apparent.

Calculations of eco-exergy storage, ascendancy, and power show that they are following the same trends when changes are made, except for the energy-based ascendancy, when an extra connection is added. While the storage and power are increased, ascendancy is decreased by the addition of an extra net flow due to the addition of ambiguity.

The eco-exergy-based ascendencies and powers are significantly higher than the energy-based ones. If the full consequences of changing a network should be calculated, for instance, in a network dynamic model (NDM) or by illustrating the evolutionary increase of ecological networks, it would therefore probably be advantageous to use eco-exergy ascendency. The experience of using eco-exergy as a goal function in structurally dynamic modeling has been positive in 23 case studies. All the case studies have, however, been on changing the properties of the key species. It cannot be excluded that major changes of the network could be covered more correctly by using the sum of eco-exergy and the eco-exergy-based ascendency as the goal function. To illustrate the evolution of the ecological network, both energy- and eco-exergy-based ascendency will be used in the next sections.

13.3 Information Embodied in the Networks and Horizontal Evolution

The networks represent not only a growth direction, but also a huge amount of information, which is crucial for the functions of ecosystems. As we have seen in Chapter 12, the networks have several important positive effects on ecosystems, illustrating that the networks contain information. The amount of information of an ecological network becomes more in accordance with the information contained in the biological components that are not in the network, if eco-exergy is used consequently to calculate the eco-exergy of both the components and the network flows.

The evolution has been both vertical, i.e., the eco-exergy/g biomass has increased, and horizontal, i.e., the number of species has increased. We exclude in this long-term view associated with the evolution the declining biodiversity due to human impact during the last couple of hundred years.

> The increase of the eco-exergy/g biomass by vertical evolution for the most complex species, from the early biological cells about 3,800 million years ago to *Homo sapiens* today, is presented in the β-value, which has increased from about 5 to 2,173 (see Table 4.1; Jørgensen et al., 2007; Jørgensen, 2008).

It would therefore be interesting to attempt to describe in a similar manner the result of the horizontal evolution. Which development of the possible ecological networks is a result of the horizontal evolution due to the increased number of possible participants in the networks, and which increase in eco-exergy will this development of the networks imply?

We do not know all the species from fossil records, because only a small fraction of the previous species have been found as fossils. On the other hand, the most dominant species are probably represented by the fossils. It may therefore, with approximation, be possible to set up typical and most probable ecological networks that have evolved due to the horizontal evolution, because we have a certain, although limited, knowledge to the dominant species, their properties, and food items. As we do not have knowledge of the ecological networks as a result of the evolution, we have to make qualified guesses, presenting possible and probable ecological networks. By calculations at different evolutionary steps of

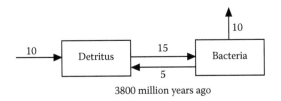

FIGURE 13.6 3,800 million years ago the earth was only inhabited by primitive cells that were living off of detritus.

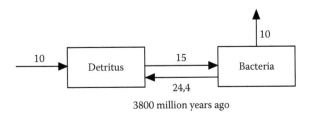

FIGURE 13.7 Energy is replaced by eco-exergy in Figure 13.6.

1. The power = the sum of energy flows
2. The eco-exergy power = the sum of eco-exergy flows
3. The ascendency, calculated based on energy flows
4. The ascendency, calculated based on eco-exergy flows

we could obtain a first rough and approximate estimation of the evolution of the ecological network. These calculations can directly be compared with the result of the vertical evolution. We may thereby be able to draw a picture of how more and more complex ecological networks due to a wider and wider spectrum of more and more advanced species have evolved. Jørgensen (2008) has made these calculations for 12 different networks representing 12 different stages of the evolution. We will not repeat all the steps in this textbook, but have chosen four typical stages and encourage the readers that want all the details to apply the reference (Jørgensen, 2008). The summarizing results of all 12 steps will, however, be given in table form. For the four steps that are given, Figures 13.6 to 13.13 show the ecological network with both energy and eco-exergy.

To focus entirely on the development of the networks, it is presumed that the ecological networks always capture the same amount of energy from the solar radiation. The idea is to illustrate the differences in the structure of the networks—not in their ability to capture solar radiation. Figures 13.6 and 13.7 show a network 3,800 million years ago, when the first primitive cells were living off detritus, probably resulting from various forms of energy, including solar radiation.

Figures 13.8 and 13.9 show a probable network 1,500 million years ago when photosynthesizing diatoms and amoebae emerged. Figures 13.10 and 13.11 illustrate a probable network 480 million years ago, 45 million years after the so-called Cambrian explosion, where many new species emerged and populated the earth.

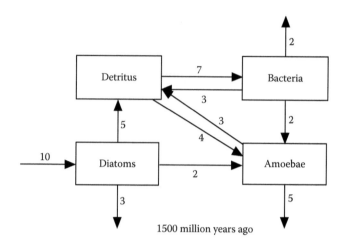

FIGURE 13.8 The cyanobacteria that emerged more than 2,300 million years ago were replaced 1,500 million years ago by diatoms, which could also photosynthesize, but are eukaryote cells. The amoeba has been introduced and the ecological network consists now of four interacting components.

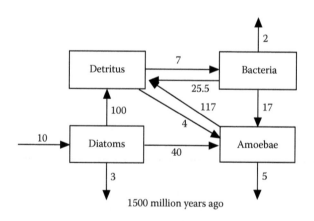

FIGURE 13.9 Eco-exergy has replaced energy in Figure 13.8.

Figures 13.12 and 13.13 represent possible and probable networks 35 million years ago, after the dinosaurs were extinguished, probably by the impact of an asteroid 65 million years ago, and the mammals had conquered the earth. These two figures, as Figures 13.6 and 13.7, show only possible networks of the main flows.

The results of calculating energy power, eco-exergy power, energy-based ascendency, and eco-exergy-based ascendency for the networks referred to in Jørgensen (2008) are summarized in Table 13.3. Only eight of the networks—four based on energy and four based on eco-exergy—have been shown in this section in Figures 13.6 to 13.13.

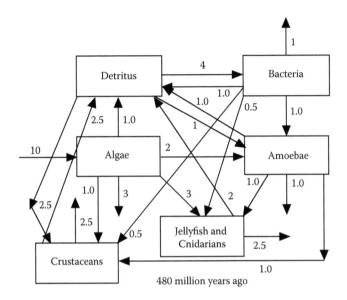

FIGURE 13.10 A wide spectrum of crustaceans has merged as a result of the Cambrian explosion.

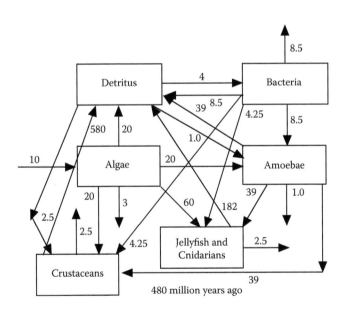

FIGURE 13.11 Eco-exergy has replaced energy in Figure 13.10. The crustaceans have a significantly higher β-value than the other organisms in the diagram, which of course implies that the eco-exergy flows are increased considerably compared with the previous diagrams.

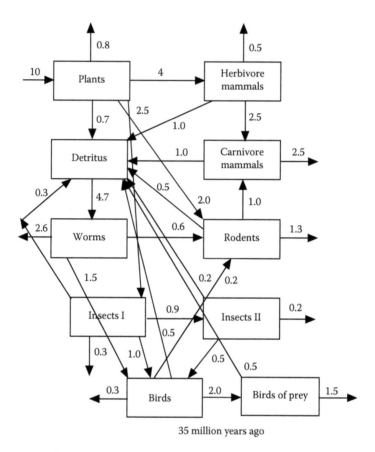

FIGURE 13.12 Three types of mammals are included in the ecological network shown: herbivore mammals, carnivore mammals, and rodents. The ecological network is probably not more complex than about 100 million years earlier, when the dinosaurs were governing the earth. The evolution from the dinosaurs to mammals has, to a high extent, been vertical, which can be seen by the β-value that has been more than doubled from dinosaurs to mammals. The same is probably the case for the evolution from the wide spectrum of mammals to man.

The energy ascendency approximately doubled during the evolution from the primitive prokaryote to the mammals. It may therefore be possible to conclude that the evolution has had a tendency to increase slightly the organization of the networks expressed by the ascendency, and the ambiguity about where a quantum in any compartment will flow next is decreased correspondingly. The energy power is roughly not increased or increased very little. The eco-exergy power and ascendency have increased 300–400 times, which can be explained by the increase of eco-exergy of the components (see Table 13.3). The eco-exergy per g of biomass has increased in the same period from 91 kJ/g to 39,775 kJ/g, or 437 times. The increase of biodiversity, as we have discussed in Chapter 10, can therefore not have caused an increase of the eco-exergy ascendency, but the increase seems entirely to be due to the increase in eco-exergy per g of biomass. These results are consistent with the results presented in Chapter 7.

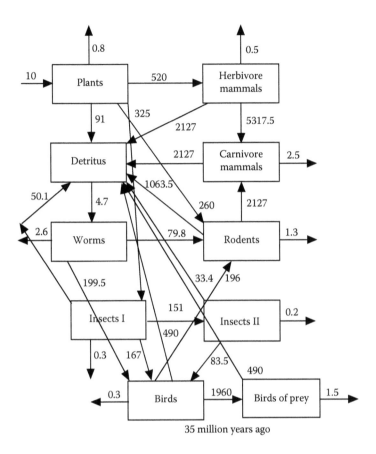

FIGURE 13.13 Eco-exergy has replaced energy in the figure. The high β-value of mammals implies that the eco-exergy flows and the eco-exergy ascendency are considerably higher than the energy flows and energy ascendency, or expressed differently, the evolution from 150 million years ago to 35 million years ago has to a high extent been vertical. It is also probably the case for the evolution from general mammals to man, i.e., the evolution of the last 35 million years.

The networks have, however, increased in complexity, and the same amount of energy—10 units—can support more and more biomass and eco-exergy, because the growth of the network implies that the mass, energy, and eco-exergy due to cycling can be utilized better and better. Table 13.4 has found the amount of biomass expressed as kJ (1 g organic matter has 18.7 kJ of energy) that the network can sustain, presuming that the network is donor based, where the outflows to the other components are all 20% of the donor's energy content. These are the same conditions used in Figure 12.3. Similarly, the eco-exergy sustained by the various networks has been found. The energy content of the various components has been multiplied by the β-values (see Table 4.1). The results of these calculations are included in Table 13.4.

The results show that the network 35 million years ago could sustain 55% more biomass, which must be due entirely to a more effective use of the incoming energy by the

TABLE 13.3　Evolution of Networks

10⁶ Years Ago	Energy Power	Eco-Exergy Power	Energy Ascendency	Eco-Exergy Ascendency
3,800	40.0	59.4	48.7	52.7
3,200	40.0	77.5	48.7	57.7
2,300	40.0	115	60.0	96.7
1,500	46.0	331	62.2	161.5
750	46.0	631	65.4	495
525	44.0	699	65.1	519
480	45.0	1,052	62.9	675
450	47.5	1,911	69.3	1,555
330	52.5	3,409	86.0	3,510
150	44.1	6,361	86.2	7,606
100	46.1	7,995	97.6	12,174
35	48.1	17,883	102	20,845

Note: The networks 3,800, 1,500, 480, and 35 million years ago have been illustrated in this section. For the other networks see Jørgensen (2008).

TABLE 13.4　Evolution of Networks

10⁶ Years Ago	Biomass Sustained by the Networks	Eco-Exergy Sustained by the Networks
3,800	100	220
3,200	100	288
2,300	100	763
1,500	130	2,045
750	120	3,050
525	120	3,092
480	125	5,218
450	132	9,677
330	135	16,412
150	121	38,430
100	137	44,090
35	155	84,737

network. The eco-exergy of the network is increased about 400 times, or about the same as the eco-exergy per g. However, the network consists of many different organisms with different eco-exergy per gram.

The development of the networks has provided the entire network with an increase in the information based on application of eco-exergy that corresponds to the most developed organisms. The development of the networks has made it possible for the entire ecosystem to increase its eco-exergy content or capital to the same level— about 400 times—as the most developed organisms, i.e., the organisms that have the most eco-exergy—the highest β-value.

13.4 Life Is Information

Ecosystem development in general is a question of the energy, matter, and information flows to and from the ecosystems and between the components of the ecosystems. No transfer of energy in ecological and biological systems is possible without matter and information, and no matter can be transferred without energy and information. The higher the levels of information, the higher the utilization of matter and energy for further development of ecosystems away from the thermodynamic equilibrium (see also Chapter 7). These three factors are intimately intertwined in the fundamental nature of complex adaptive systems such as ecosystems, in contrast to physical systems, which most often can be described completely by material and energy relations. Life is therefore both a material and a nonmaterial phenomenon. The self-organization of life essentially proceeds by exchange of information. Of the three qualities that characterize life—mass, energy, and information—it is information that defines life. Indeed, it can be argued that information does not exist in the absence of life (Eigen, 1992). We measure and model mass and energy flows in ecosystems because we can, but life is information (Jørgensen, 2008).

The information content increases in the course of ecological development because an ecosystem encompasses an integration of all the information and signals that are imposed on the environment. Thus, it is on the background of genetic information that systems develop, which allows interaction of information with the environment. Herein lies the importance in the feedback between organisms and their environment. It entails that an organism can only evolve in an evolving environment, which is modified by the organism.

The conservation laws of energy and matter set limits to the further development of "pure" energy and matter, as was made clear in the first chapters of the book, while information may be amplified (almost) without limits. Limitation by matter is known from the concept of the limiting factor: Growth continues until the element that is the least abundant relative to the needs of the organisms is used up (see Chapter 2). Very often in developed ecosystems (for instance, an old forest) the limiting elements are found almost entirely as organic compounds in the living organisms, while there are no or very few inorganic forms left in the abiotic part of the ecosystem. The energy input to ecosystems is determined by the solar radiation, and many ecosystems capture about 75–80% of the solar radiation that is the upper physical limit due mainly to the limitation given by the second law of thermodynamics. The eco-exergy of the information content of a human being can be calculated by the use of the equations in Chapters 4, 6, and 7 to be about 40 MJ/g.

A human body of about 80 kg contains about 8 kg of proteins. Let us presume randomly that only 0.06% of the proteins at most could be coded as enzymes that are controlling the life processes. One gram of biomass would then contain 100 mg of proteins or 0.06 mg of life-controlling enzymes. It could presumably represent a maximum content of information. If we presume an average molecular weight of the amino acids making up the enzymes of 200, the number of amino acid molecules would be 60 µg divided by 200 and multiplied by Avogadro's number: $3 \times 10^{-7} \times 6.2 \times 10^{23}$ = about 2×10^{17}. The corresponding β-value can be found from Equation (4.34):

$\beta = 4.00 * 10^{-6} * AMS$ (AMS = number of amino acids in the right sequence) $= 8 * 10^{11}$

This means that 1 g of biomass could have an amount of eco-exergy equal to $8 * 10^{11} * 18.7 = 1.5 * 10^{13}$ kJ, as it is presumed that 1 g of organic matter has 18.7 kJ of free energy = eco-exergy. The calculated amount of eco-exergy corresponds, in other words, to 1.5×10^{13} kJ/g body weight.

These calculations are back of the envelope calculations and do not represent what is expected for the information content of organisms in the future, but it seems possible to conclude that the development of the information content is very far from reaching its limit, in contrast to the development of the material and energy relations. See also Figure 13.14.

Information has some properties that are very different from mass and energy. It is very important to understand these differences to understand ecosystems. The properties of information that should be emphasized are:

1. *Unlike matter and energy, information can disappear without trace.* When a frog dies the enormous information content of the living frog may still be there microseconds after the death in the form of the right amino acid sequences, but the information is useless, and after a few days the organic polymer molecules have decomposed.

    It is characteristic for life that it is able to apply information. Life processes are embodied in information.

2. *Information expressed, for instance, as eco-exergy, in energy units, is not conserved.* Property 1 is included in this property, but in addition, it should be stressed that living systems are able to multiply information by copying already achieved successful information, which implies that the information survives and is utilized. The information is by autocatalysis able to provide a pattern of biochemical processes that ensure survival of the organisms under the prevailing conditions determined by the physical-chemical conditions and the other organisms present

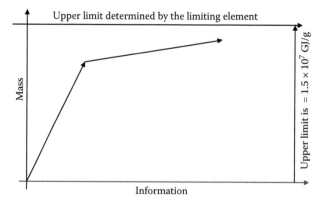

FIGURE 13.14 Growth of biomass is limited by the limiting elements or by the amount of incoming solar radiation, while the information content of living organisms is very far from its limit.

in the ecosystem. By the growth and reproduction of organisms the information embodied in the genomes is copied. Growth and reproduction require input of food, and it could be considered to include in the energy content of the food, the energy that the evolution has cost. In that case, the copy process would cost more energy than just the 18.7 kJ/g of organic matter.

3. *The disappearance and the copying of information, that are characeristic processes for living systems, are irreversible processes.* A made copy cannot be taken back, and life and death are irreversible processes. Although information can be expressed as eco-exergy in energy units, it is not possible to recover chemical energy from information on the molecular level, as known from the genomes. It would require a Maxwell demon that could sort out the molecules, and it would therefore violate the second law of thermodynamics. The role of information is directly connected with the problem of Maxwell's demon (Tiezzi, 2006). Brillouin (1962) has shown that the demon would require information about the molecules, and that more energy would be needed to obtain this information than the energy gained, which means that Maxwell's demon does not exist. There are, however, challenges to this second law (see Capek and Sheehan, 2005), and this process of copying information at very low costs could be considered one of them. Mass can be transformed by nuclear processes to energy, as expressed by Einstein: $e = mc^2$, but transformation of energy to mass is not a natural process on the earth, because it implies that matter and antimatter are formed, a process that was very common in the early universe. Energy can be transformed to information and macroscopic information to energy (see Figure 13.15), but information on the molecular level would, as already emphasized, require a Maxwell demon. Figure 13.16 illustrates the irreversibility of the transformation processes between mass, energy, and information.

It is easy to go from mass to energy to information, but it is not necessarily possible under all circumstances to go from information to energy and from energy to mass.

4. *Exchange of information is communication, and it is this that brings about the self-organization of life.* Life is an immense communication process that happens in several hierarchical levels. Exchange of information is possible without participation of mass and energy, while storage of information requires that the information is linked to material, for instance, the genetic information is stored in the genomes and transferred to the amino acid sequence.

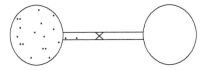

FIGURE 13.15 The left chamber contains 1 mole of a pure ideal gas, while the right chamber is empty. If we open the valve, the system will lose eco-exergy (or technological exergy) = RT ln 2, which we could utilize by installation of a propeller in the valve. The entropy of the system will simultaneously increase by R ln 2. In this case, macroscopic information is converted to eco-exergy.

FIGURE 13.16   Matter can be transformed to energy and energy to information, which can be copied at low costs. Transformation of information to energy requires that it is macroscopic information (see Figure 1.5), and transformation of energy to mass implies that both matter and antimatter are formed. The dotted arrows indicate that the transformation is only possible under certain conditions.

13.5 Summary of the Important Points in Chapter 13

1. The individuals in an ecosystem, as the ecological network, embody an enormous amount of information.
2. The network information can be expressed by the ascendency, which is defined in Figure 13.1. It is beneficial to apply the flows as eco-exergy flows by calculation of the ascendency to obtain an accordance between the information embodied in the genomes and that in the networks.
3. The increase of the eco-exergy/g biomass by vertical evolution for the most complex species, from the early biological cells about 3,800 million years ago to *Homo sapiens* today, is presented by the β-value, which has increased from about 5 to 2,173.
4. The increase of biodiversity is the horizontal evolution, which implies that more and more complex networks and more and more effective networks have been developed.
5. The development of the networks has made it possible for the entire ecosystem to increase its eco-exergy content or capital to the same level—about 400 times—as the most developed organisms, i.e., the organisms that have the most eco-exergy, meaning the highest β-value.
6. Information:
   a. Is very far from its limits
   b. Is embodied in life processes and life is information
   c. Is not conserved
   d. Is copied and deleted by irreversible processes
   e. Is exchanged by communication, and it is this that is bringing about the self-organization of life

Exercises/Problems

1. Explain the relation between the ambiguity of the network and the network connectivity.
2. Calculate in detail, step-wise, the ascendency of the network in Figure 12.7 with both energy and eco-exergy.
3. The information embodied in the genes is far from its limits. Is the information embodied in the ecological network close or far from its limits? Give a detailed explanation of the answer.

4. Give an example from society that illustrates the properties of information, particularly the properties C, D, and E under point 6 in the summary.
5. Give the difference between macroscopic and microscopic (molecular) information. Explain why it is impossible to transfer the microscopic information to free energy opposite the macroscopic information.
6. Give an example of macroscopic and an example of microscopic information.

14

Ecosystems Have Emerging Holistic System Properties

A system is more than sum of its parts.

The properties of ecosystems cannot be explained by the components alone. Ecosystems are much more than the sum of the components. They have unique holistic properties that explain how they can grow and develop according to the ecological law of thermodynamics (ELT), and in spite of that they are subjugated to the thermodynamic laws and the biochemistry on earth. The ecosystem properties have been presented in Chapters 8–13. They are rooted in the development of a high and effective utilization of the three growth forms: growth of biomass, information, and networks. These properties are integrated by the use of a holistic system approach.

14.1 Introduction

A human body consists of a number of chemical compounds: lipids, proteins, calcium compounds, and so on. The total value of these chemical compounds is maybe on the order of $100, at the most. So, it is not the value of the chemical compounds that makes the human being (or any other mammal) a unique creature, or we may say system. It is the cooperation of these compounds in various synergistic networks based on an enormous organizational knowledge that makes it possible for a human being to coordinate and direct his movements, to come up with new ideas, solve problems, speak, sense, read, write, and express emotions. A human being is much more than just a collection of chemical compounds. He is a self-organizing system with many surprising and advantageous properties.

The ecosystems work similarly as unique systems with surprising and advantageous properties, which have been presented in the previous chapters. Ecosystems are like all of nature subjugated to the thermodynamic laws and biochemistry. They are a result of a selection made by early life and which was already characteristic for the first prokaryote and later for the eukaryote cells. All systems are subjugated to the second law of thermo-dynamics, which implies that it costs free energy to maintain a system more or less far from the thermodynamic equilibrium, but if an inflow of free energy is able to deliver the required maintenance free energy, it is of course possible to maintain the system away from the thermodynamic equilibrium. If the free energy flow is more than the needed maintenance free energy, it is even possible for the system to move further away from thermodynamic equilibrium. Ecosystems are characterized by having a huge inflow of free energy from a long-term stable source—the sun. It has furthermore been shown that if the free energy flow is more than the maintenance free energy, cycling of matter will inevitably get started in the system (Morowitz, 1968). As discussed in Chapters 6 and 7, life offers many possibilities to move away from thermodynamic equilibrium. Darwin's theory claims that among the many possibilities, the one corresponding to the properties of the components that give most survival will win, and translated to thermodynam-ics, this means that the "solution" that moves the system furthest from thermodynamic equilibrium will win. Darwin's theory focuses on the species, while the ELT in Chapter 7 integrates the results of Darwin's theory applied on all species in an ecosystem, which is possible because the species are cooperating and interacting in networks and in a well-organized hierarchical system. The distance from thermodynamic equilibrium is deter-mined by the eco-exergy according to the definition of this concept, and we can therefore, as we have done in Chapters 6 and 7, formulate what is denoted the ecological law of thermodynamics (ELT), or sometimes the fourth law of thermodynamics, as follows:

> The ecological law of thermodynamics (ELT): A system that receives a throughflow of exergy (high-quality energy, free energy, energy that can do work) will try to utilize the free energy flow to move away from thermodynamic equilibrium, and if more combinations of components and processes are offered to utilize the exergy flow, the system will select the organization that gives the system as much exergy content (storage) as possible, i.e., maximizes dEx/dt.

The ability of ecosystems to move away from thermodynamic equilibrium has been developed throughout the entire evolution. The result is that ecosystems today have unique properties that enable them to utilize the free energy inflow extremely well. Components containing more and more information have been developed, which means that not only do the components contain more eco-exergy (as information has eco-exergy), but the information tells the components how they can utilize the flow of free energy better to move further away from thermodynamic equilibrium.

Chapters 8–13 have demonstrated how ecosystems have unique and very advanta-geous properties that allow them to utilize the free energy flow from the solar radiation to move further away from thermodynamic equilibrium. "Mother nature" has really given the ecosystems during the evolution better and better properties to enable them to gain more and more eco-exergy—to move further away from thermodynamic equilib-rium. This process has continued for a very long time, which means that "father time" has also played an important role. Ecosystems have utilized, as already mentioned, three growth forms to move away from thermodynamic equilibrium: growth of biomass, growth of information, and growth of the ecological network.

Chapters 8–13 present the holistic properties of ecosystems that have enabled them to utilize the three growth forms. Ecosystems are of course nonisolated; otherwise, they could not receive the free energy flow, that is, the prerequisite for all three growth forms. The openness of ecosystems explains that ecosystems can maintain a very complex structure that is far from thermodynamic equilibrium. Without the inflow of free energy, the ecosystems could not use the first growth form—to build biomass on the basis of solar radiation by photosynthesis up to the point where the most limiting element has been used up. Openness of ecosystems implies furthermore that the ecosystems can receive new information—new possibilities for growth, for instance, by immigration. The hierarchical organization is, however, also important for the possibilities of ecosystems to move away from thermodynamic equilibrium by a better utilization of the inflowing free energy by solar radiation. Ecosystems are hierarchically organized in such a way that the openness, and therefore the inflow of free energy on the different hierarchical levels, is coordinated to the dynamic of the level (see Chapters 8 and 9). Also, the information content in the various levels of the ecological hierarchy complies with the function of the level. The cells, for instance, use enzymes that are based on the genetic information to direct the life processes. A selection of the best processes and the most effective enzymes for the direction of these processes currently takes place. The species, on the other hand, have important macroscopic properties that allow them to survive and even gain more eco-exergy, and their properties are also steadily improved by selection.

The information content is central, and therefore an ecosystem can move away from thermodynamic equilibrium by increasing the information—one of the three growth forms—which can continue to grow almost infinitely opposite the growth of biomass, which is limited by one or more needed elements. The diversity of ecosystems is enormous on all levels of the ecological hierarchy, as presented in Chapter 10. The high diversity furthermore has the advantage that it facilitates finding for the ecosystem a solution to resist new disturbances that the ecosystems have not experienced before (see Chapter 10). The information furthermore implies that ecosystems possess self-organization that enables them to resist disturbances and recover afterwards, as discussed in Chapter 11.

The formation of ecological networks is very important for ecosystems and their utilization of the free energy flow, as presented in Chapter 12. Development of the ecological networks is furthermore, as already outlined, a possibility for ecosystems to gain eco-exergy and to move further away from thermodynamic equilibrium by the information embodied in the network and by a higher efficiency of the available free energy for the ecosystem. Information embodied in all levels of the ecological hierarchy and in the ecological network unites to a certain extent holistically all three growth forms and their properties—openness, hierarchical organization, high diversity, high buffer capacity, and a very well-developed network, which all contribute to enhancing the possibilities to move away from thermodynamic equilibrium. In short, we may express it as *life is information*.

The properties presented in Chapters 8–13 explain why and how ecosystems are able to follow the ecological law of thermodynamics (ELT). To understand these holistic properties is important if we want to understand how ecosystems work and how they are able to move away from thermodynamic equilibrium in spite of being subjugated to the thermodynamic laws and the biochemistry characteristics for the earth. Ecosystems are the most unique systems in the universe, because of their self-organizing ability, their enormous complexity, their continuous irreversible evolution, and their enormous content of information (see also Odum, 1989, 1996). Systems ecology is about these unique systems, their holistic properties, and the conditions that the natural laws have imposed on these systems.

In the next section, we will present several holistic properties of ecosystems that an observer would notice in addition to the properties that explain that ecosystems can obey ELT under the basic conditions dictated by nature. We could in principle refer to any textbook in ecology to present a list of these properties and prefer to apply the results by observers applying ecosystem ecology.

Example 14.1

The industries in the Danish town Kalundborg are cooperating in a network and thereby form an industrial system. Below the network is shown. Indicate three properties of this industrial system that comply with the properties of ecosystems and three properties that do not comply with the properties of ecosystems. In this context, it should be mentioned that agricultural systems can be built in networks. The integrated agriculture, which was characteristic 70–80 years ago in the developed countries, was organized similarly to the industrial system shown and with approximately the same benefits and disadvantages as this system.

The industrial network in the Danish town Kalundborg is shown. The connections are flows of matter; see the key for the type of flows.

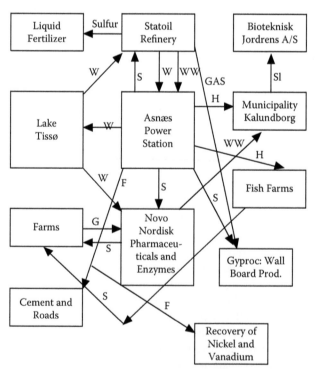

W: Water, F: Flyash, WW: Waste water, S: Steam, Sl: Sludge, H: Heat, G: Grains

Solution

Properties similar to ecosystems:

1. The cooperation in the network implies increased efficiency in the use of the resources. Some components use the waste from other components as raw material.
2. The synergism and mutualism (particularly in economic terms) have increased due to the cooperation in the network.
3. The industrial system has more structure, better economy, and more information than the components would have together if they were not cooperating in a network. They have moved further away from thermodynamic equilibrium by the cooperation in a network.

Properties not similar to ecosystems:

1. A complete recycling is not taking place.
2. The organization is not completely hierarchical.
3. The diversity is much less than for ecosystems.

14.2 Additional Properties of Ecosystems

Nielsen (2007) and Nielsen and Mueller (2009) have, as observers, listed and discussed several ecosystem properties that they applied in industrial ecology and for the presentation of ecosystem services. They have listed the following holistic properties of ecosystems:

1. Complexity—components and interconnectedness
2. Evolution—adaptation and selection
3. Compartmentalization—quantitative and qualitative perspectives
4. Flows and processes—quantitative and qualitative perspectives
5. Cybernetics—feedbacks and controls, quantitative and qualitative perspectives
6. Cycling—cycles, cycling, and recycling
7. Network properties—synergism and utility
8. Organizational issues
9. Diversity
10. Openness and dissipation

A few comments and reference to the basic properties presented in Chapters 8–13 are given below.

1. Complexity. It is of course a consequence of the information content and the diversity of all the levels in all the ecological hierarchy. The complexity of ecosystems is determined by their openness far from thermodynamic equilibrium. It is clear from Chapter 9 that ecosystems have an enormous complexity, due not only to the many different components on all the hierarchical levels, but also due to the many flows and processes that are continuously operating in all ecosystems. Ecosystems are dynamic systems due to these flows and processes, which are functions of time, as all the components in all the hierarchical levels are also functions of time.

Living systems as ecosystems are many magnitudes more complex than physical systems.

2. Evolution. It is a consequence of ELT, as the components of the ecosystems are struggling under the prevailing conditions to move as far as possible away from thermodynamic equilibrium. At the same time, the prevailing conditions determined by the forcing functions, which vary in time and space, are continuously changing. The ecosystems generate simultaneously new possibilities to meet the prevailing conditions on all levels of the ecological hierarchy. Of particular importance is the generation of new possibilities on the species level, by mutations, sexual recombinations, and general transfer of genes among organisms. Ecosystems evolve on all the levels of the ecological hierarchy, and the evolution is both vertical and horizontal, which implies that the information embodied in the genes is increasing and the complexity and efficiency of the networks are increasing.

3. Compartmentalization. The description of the hierarchical organization requires that on each level we have different compartments, which of course have clear definitions and boundaries. The compartmentalization is already introduced in the various scientific disciplines that we are using in this context: elementary particles, atoms, molecules, cells, organs, species, populations, communities, ecosystems, landscapes, regions, and the ecosphere (which is part of a planet that is carrying life).

4. Flow and processes. These are the transfers among the compartments in the network. It is clear that a more quantitative description of ecosystems, for instance, in ecological models, will require that mathematical expressions are applied for the various flows and processes. It should be emphasized that ecological networks consist of not only compartments that are connected in a network, but also connections that are flows and processes with a causality, which we want to capture when we are constructing ecological models.

5. Cybernetics, feedback, and controls. Ecosystems possess an enormous self-organization and self-regulation. All processes in an ecosystem are practically regulated by enzymes directly or indirectly. The concentrations of the enzymes and the conditions of the processes, such as pH and temperature, are to a great extent also regulated by processes. Because of the wide spectrum of possibilities and components on all levels of the ecological hierarchy and the continuous selection, there is an enormous ability to adapt to the prevailing conditions.

6. Cycling. Matter, all the biologically important elements, energy, and information cycle in ecosystems, which is the immediate result of the network. The cycling is crucial for ecosystems, because all further growth would stop when the most limiting element was used up in the form applied, if the elements were not returned or recovered by the cycling. Energy cycling is important because it makes it possible, as demonstrated in Chapter 12, to increase the efficiency of the free energy utilization. Information cycling is important because a dissemination of the information to all corners of the ecosystem is thereby made possible, and it can be done by the network at no cost or almost no cost. Without the cycling, no evolution would have taken place, because as already stressed, the growth would stop and the struggling for growth is the prerequisite for the evolution.

7. Network properties have been presented in detail in Chapter 12. They are crucial for the ability of ecosystems to follow and utilize the ELT. It is also understandable from Chapter 12 that growth of networks as one of the three growth forms is significant for ecosystems and evolution.

8. Organization. The hierarchical organization is presented in Chapter 9, and it is clear from the details given in this chapter that this organization is very beneficial for the ecosystems—not only because the openness of the different levels makes it possible to obtain the right dynamics on all levels, but also because disturbances are absorbed very effectively. Furthermore, the diversity on the different levels fits the size, processes, dynamics, and components of the levels.

9. Diversity. In light of the enormous variability of the forcing functions in time and space, it is very beneficial for ecosystems to meet this variability of the challenges to move as far away as possible from thermodynamic equilibrium by the availability of as many solutions as possible. It is clearly presented in Chapter 10 that diversity on all the ecological hierarchical levels is urgently needed to meet all possible challenges and disturbances. There seems to be a correlation between PET (the amount of water that could be evaporated from the soil and transpired by the plants given the average temperature and humidity; PET expresses the energy input to the ecosystem) and the species diversity. Higher PET means that larger populations could be supported by the ecosystems, implying a reduced probability of extinction. It would be a method that the ecosystem could apply to maintain a particularly high eco-exergy by the favorable conditions of a high PET.

10. Openness and dissipation. Because of the dissipation of free energy, ecosystems must be open or at least nonisolated. The huge complexity of ecosystems makes ecosystems ontic open in the sense that it is impossible to make certain predictions, as is possible for physical systems. It is, however (and what this book is about), possible to give high probabilities or propensities for possible changes of ecosystems as a result of given changes of forcing functions or impacts on ecosystems.

14.3 Summary of Important Points in Chapter 14

1. Ecosystems cannot be considered a collection of molecules, cells, or species, but are systems with unique, holistic, self-organizing, and self-regulating properties that make it possible for the ecosystems to follow ELT, in spite of the fact that they are subjugated to the thermodynamic laws and the biochemistry of the earth.

2. The holistic properties of ecosystems have made it possible for ecosystems to move further and further away from thermodynamic equilibrium by favoring the utilization of the three growth forms: growth of biomass, information, and networks.

3. The holistic system properties of ecosystems are all rooted in information: life is information.

4. Important ecosystem properties to understand the reactions and processes of ecosystems are:
   a. Complexity—components and interconnectedness

   b. Evolution—adaptation and selection
   c. Compartmentalization—quantitative and qualitative perspectives
   d. Flows and processes—quantitative and qualitative perspectives
   e. Cybernetics—feedbacks and controls, quantitative and qualitative perspectives
   f. Cycling—cycles, cycling, and recycling
   g. Network properties—synergism and utility
   h. Organizational issues
   i. Diversity
   j. Openness and dissipation

Exercises/Problems

1. Give at least three examples of other systems than ecosystems that possess emergent, holistic system properties.
2. A town is a system. Give a short overview of the similarities and differences between towns and ecosystems.
3. Give at least three examples of the ecosystem properties that we unfortunately are missing in all systems built by humans, for instance, towns, society, or a country.

15

Application of System Ecology in Ecological Subdisciplines and Environmental Management

A vision without action is just a dream; an action without vision just passes time; a vision with action changes the world.

—Nelson Mandela

Systems ecology is applied widely in environmental management and particularly in the ecological subdisciplines, which are the fundaments for ecological, holistic, and integrated environmental management: ecological modeling, ecological engineering, and assessment of ecosystem health or integrity by ecological indicators. These important applications of systems ecology are presented in this chapter.

15.1 Integrated Ecological and Environmental Management Should Be Based on a Profound Knowledge of System Ecology

Integrated ecological and environmental management means that the environmental problems are viewed from a holistic angle considering the ecosystem as an entity and considering the entire spectrum of solutions, including all possible combinations of proposed solutions. The experience gained from environmental management the last 40 years has clearly shown that it is important not to consider solutions of single problems, but to consider *all* the problems associated with a considered ecosystem simultanously

269

applying a system view and evaluate *all* the solution possibilities proposed by the relevant disciplines at the same time, or expressed differently: to observe the forest and not the single trees. The experience has clearly underlined that there is no alternative to an *integrated* management, at least not on a long-term basis. Fortunately, new ecological subdisciplines have emerged, and they offer toolboxes to perform an integrated ecological and environmental management. As well, the application of these toolboxes and environmental management in general are based on a solid fundament of knowledge about the core systems—the ecosystems, which implies that there is a need for system ecology.

Integrated ecological and environmental management consists of a procedure with seven steps (see Jørgensen and Nielsen, in press):

1. Define the problem.
2. Determine the ecosystems involved.
3. Find and quantify all the sources to the problem.
4. Set up a diagnosis to understand the relation between the problem and the sources.
5. Determine all the tools we need to implement to solve the problem.
6. Implement the selected solutions.
7. Follow the recovery process.

When an environmental problem has been detected, it is necessary to determine and quantify the problem and all the sources to the problem. It requires the use of analytical methods or a monitoring program. To solve the problem a clear diagnosis has to be developed: What are actually the problems the ecosystems are facing, and what are the relationships between the sources and their quantities and the determined problems? Or expressed differently: To what extent do we solve the problems by reducing or eliminating the different sources of the problems? A holistic integrated approach is needed in most cases because the problems and the corresponding ecological changes in the ecosystems are most often very complex, particularly when several environmental problems are interacting. This step therefore requires system ecology. When the first green wave started in the mid-1960s, the tools to answer these questions, which we today consider very obvious questions in an environmental management context, were not yet developed. Meanwhile, systems ecology has been developed. We could 40 years ago carry out the first three points, but on a weaker basis, and we had to stop at point 4 and could at that time only recommend to eliminate the source completely or close to completely by the available methods—by environmental technology. New and improved environmental technological methods have of course been developed during the last 45 years.

Due to the development of several new ecological subdisciplines, it is today possible to accomplish points 4–7. They are the result of the emergence of six, or two times three, new ecological subdisciplines during the last 40 years. For a better diagnosis, ecological modeling, ecological indicators, and ecological services are applied.

More tools to solve the problems are ecological engineering (also denoted ecotechnology), cleaner production, and environmental legislation. In parallel, systems ecology has developed hand in hand to a certain extent with the six new subdisciplines. These new subdisciplines will be presented below, together with the need for systems ecology for the development and applications of these subdisciplines.

A massive use of ecological models as an environmental management tool was initiated in the early 1970s. The idea was to answer the question: What is the relationship between a reduction of the impacts on ecosystems and the observable, ecological improvements. The answer could be used to select the pollution reduction that the society would require and could afford economically. Ecological models were actually already developed in the 1920s by Steeter-Phelps and Lotka Volterra (see, for instance, Jørgensen and Fath, 2011), but in the 1970s a much more consequent use of ecological models started, and many more models of different ecosystems and different pollution problems were developed. Today we have at least a few models available for almost all combinations of ecosystems and environmental problems. The journal *Ecological Modelling* was launched in 1975 with an annual publication of 320 pages and about 20 papers. Today, the journal publishes 20 times as many papers. This means that ecological modeling has been adopted as a very powerful tool in ecological-environmental management to cover particularly point 4 in the integrated ecological and environmental management procedure proposed above. Ecological models are powerful management tools, but they are not easily developed. They require in most cases good data, which are resource and time-consuming to provide. They require also a good knowledge of the systems, which are models—meaning system ecology. During the last 35–40 years ecological modeling has drawn upon system ecology and been used as a research tool to gain more knowledge about ecosystems, their processes, and reactions.

About 20 years ago, another ecological subdiscipline applicable for carrying out step 4 was proposed: ecological indicators (see, for instance, Costanza et al., 1992). Ecological indicators can be classified according to the spectrum from a more detailed or reductionistic view to a system or holistic view (see Jørgensen, 2002). The reductionistic indicators can, for instance, be a chemical compound that causes pollution or specific species. A holistic indicator could, for instance, be a thermodynamic variable or the biodiversity. The indicators can either be measured or determined by the use of a model. In the latter case, the time consumption is of course not reduced by the use of indicators instead of models, but the models get a clearer focus on one or more specific state variables, namely, the selected indicator, which best describes the problems. In addition, indicators are usually associated with very clear and specific health problems of the ecosystems, which of course is beneficial in environmental management. Again, the selection of ecological indicators, particularly the holistic indicators, requires a good knowledge of ecosystems.

Since the late 1990s, the services offered by the ecosystems to the society have been discussed, and it has been attempted to calculate the economic values of these services (Costanza et al., 1997). A diagnosis that would focus on the services actually reduced or eliminated due to environmental problems could be developed. Another possibility of using ecological services to assess the environmental problems and their consequences could be to determine the economic values of the overall ecological services offered by the ecosystems and then compare those with what is normal for the type of ecosystems considered. Jørgensen (2010) has determined the values of all the services offered by various ecosystems by the use of the ecological holistic indicator eco-exergy expressing the total work capacity (see Chapter 9). It is a good measure of the total amount of ecological services, as all services require a certain amount of free energy, i.e., energy that can do work. Assessment of the values of the ecosystem services may also be coupled to sustainability, because it is crucial to maintain the many ecosystem services on which society is dependent. Assessment of the ecosystem services frequently uses ecological indicators, but it is under all circumstances not possible to quantify and understand the services offered by ecosystems without a good knowledge of system ecology.

The toolbox environmental technology was the only methodological discipline available to solve the environmental problems 45 years ago when the first green waves started in the mid-1960s. This toolbox was only able to solve the problems of point sources, but sometimes at a very high cost. Today fortunately, environmental technology has developed significantly during the last 45 years, and we now have additional toolboxes that can solve the problems of the diffuse pollution or find alternative solutions at lower costs when the environmental technology would be too expensive to apply. As for the diagnostic toolboxes, these toolboxes are developed on a basis of new ecological subdisciplines and supported strongly by the development of system ecology.

To solve environmental problems we have today four toolboxes:

1. Environmental technology
2. Ecological engineering, also denoted ecotechnology
3. Cleaner production, and under this heading we would also in this context include industrial ecology
4. Environmental legislation

Environmental technology was available by the emergence of the first green waves about 45 years ago. Since then, several new environmental-technological methods have been developed, and all the methods have been streamlined and are generally less expensive to apply today. There is and has been, however, an urgent need for other alternative methods to be able to solve the entire spectrum of environmental problems at an acceptable cost. The environmental management today is more complicated than it was 45 years ago because of the many more toolboxes that should be applied to find the optimal solution, and because global and regional environmental problems have emerged. Figure 15.1 shows the complexity of environmental management today due to the bigger choice of toolboxes and the emergence of regional and global problems.

The toolbox containing ecological engineering methods has been developed since the late 1970s. Ecological engineering is defined as a design of sustainable ecosystems that integrate human society with its natural environment to the benefit of both (Mitsch and Jørgensen, 2004). It is an engineering discipline that operates in ecosystems, which implies that it is based on both design principles and ecology. Obviously, ecological engineering is drawing very much on ecology and systems ecology, as it is engineering operating in ecosystems which inevitably requires that the characteristics and features of the ecosystems are known. It can clearly be seen in the book on ecological engineering by Mitsch and Jørgensen (2004) where 19 principles for the proper application of ecological engineering are presented. These principles are clearly rooted in systems ecology. The toolbox contains four classes of tools:

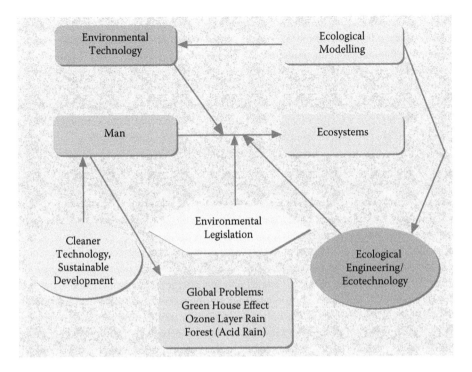

FIGURE 15.1 (See color insert.) Conceptual diagram of the complex ecological-environmental management of today, where there are various toolboxes available to solve the problems and where the problems are local, regional, and global.

1. Tools that are based on the use of natural ecosystems to solve environmental problems (for instance, the use of wetlands to treat agricultural drainage water)
2. Tools that are based on imitation of natural ecosystems (for instance, construction of wetlands to treat wastewater)
3. Tools that are applied to restore ecosystems (for instance, restoration of lakes by the use of biomanipulation)
4. Ecological planning of the landscape (for instance, the use of agroforestry)

The introduction of ecological engineering has made it possible to solve many problems that environmental technology could not solve, first of all, nonpoint pollution problems and a fast restoration of deteriorated ecosystems.

Some environmental problems can, however, not be solved without a more strict environmental legislation, and for some problems a global agreement may be needed to achieve a proper solution, for instance, by out-phasing the use of Freon to stop or reduce the destruction of the ozone layer. Notice that environmental legislation also requires an ecological insight to assess the required reduction of the emission that is needed through introduction of the environmental legislation.

As environmental legislation has been tightened, it has become more and more expensive to treat industrial emissions, and the industry has, of course, considered whether it was possible to reduce the emission by other methods at a lower cost. That has led to development of what is called cleaner production, which means producing the same product by a new method that would give a reduced emission, and therefore less cost for the pollution treatment. New production methods have been developed by the use of innovative technology, which has created completely new methods to produce the same products with fewer environmental problems. Other emission reductions have been developed by the use of ecological principles on the industrial processes, for instance, recycling and reuse. In many cases it has also been possible to achieve a reduction of the environmental problems by identification of unnecessary waste. Industrial ecology could, in the author's opinion, be defined as the use of the systems ecological principle in production, such as recycling, reuse, and holistic solutions to achieve a high efficiency in the general use of the resources.

Today, we have in the four toolboxes for environmental management solutions a possibility to solve any environmental problem, and often at a moderate cost, and sometimes even at a cost that makes it beneficial to solve the problem properly. As is the case for the diagnostic toolboxes, the toolboxes with problem solution tools are also rooted in recently developed ecological subdisciplines, which are named after the tools: ecological engineering, environmental legislation, and cleaner technology. The developments of these ecological subdisciplines, which are the basis for the application of the presented toolboxes, are all working very closely together with systems ecology. It is therefore not surprising that all the new ecological subdisciplines and systems ecology mutually have fertilized the development of each other.

15.2 The Application of Systems Ecology to Explain Ecological Observations and Rules

In Chapter 7, nine ecological rules based on observations were presented to support the ecological law of thermodynamics (ELT). Furthermore, structurally dynamic modeling has been presented, as additional support for ELT, because these types of models use eco-exergy as a goal function or orientor, and this model type has been able to explain the observed structural changes in 23 cases. ELT is, as also discussed in Chapter 7, in accordance with the observed vertical and horizontal evolution and the evolutionary theories. Systems ecology could be used much more widely to explain ecological observations, which of course would be very beneficial in environmental management, as it would make it possible to apply systems ecology to predict changes in ecosystems as a result of changes introduced by man-controlled forcing functions. The use of theory to determine the expected changes as a consequence of our intervention is of course always much more cost moderate than observations or experiments, which are often very expensive to provide. In this section, we will present further examples of ecological observations and rules that can be explained by systems ecology—often by ELT (see further details in Jørgensen et al., 2007).

1. Every garden owner knows how difficult it is to control weeds. The reason is that in a garden only a limited number of species are represented, and nature would like to use all available species to obtain the highest possible eco-exergy per ha. Leave your garden to be conquered by weeds, and a determination of the eco-exergy per ha will show that it is much more than in the weed-controlled garden.

2. In Example 7.3 Cope's rule—species have a tendency to develop toward bigger organisms' size—was shown valid for horses.

3. In his famous book, Wallace (1889) devoted a chapter to the topic "warning coloration and mimicry with special reference to the Lepidoptera." One of the most conspicuous day-flying moths in the eastern tropics was the widely distributed species *Ophthalmis lincea* (Agaristidae). These brightly colored moths have developed chemical repellents that make them distasteful, saving them from predation (Kettlewell, 1965).

4. Wallace (1858) proposed that insects that resemble in color the trunks on which they reside will survive the longest, due to concealment from predators. The relatively rapid rise and fall in the frequency of mutation-based melanism in populations that occurred in parallel on two continents (Europe and North America) is a convincing example for rapid microevolution in nature caused by mutation and natural selection. The hypothesis that birds were selectively eating conspicuous insects in habitats modified by industrial fallout is consistent with the data (Majerus, 1998; Cook, 2000; Coyne, 2002; Grant, 2002).

5. It has been found that the parasites often use the strategy to have a high initial mortality, which is decreasing. We should therefore expect that a high decreasing mortality gives a higher eco-exergy than a constant or increasing mortality, if we could use ELT to explain ecological observations. The results obtained by the use of a parasite-bird model (eco-exergy is indicated in GJ) are presented in Table 15.1. The model applied is structurally dynamic; see also Chapter 7.

   As seen, a decreasing mortality from 1.00 to 0.25 gives the highest eco-exergy totally and for the birds (which of course survive better when the parasites have a high initial mortality), but even for the parasites, because they have better conditions when more birds have survived. An increasing mortality clearly gives lower eco-exergy, and so does a constant mortality corresponding to the average of the decreasing or increasing mortality. A lower constant mortality of the parasites (0.25) gives a higher eco-exergy of parasites and a lower of the birds, but a significant lower eco-exergy totally.

6. On a global scale, species diversity typically declines with increasing latitude toward the poles (Rosenzweig, 1996; Stevens and Willig, 2002). The determinant

TABLE 15.1 Results Obtained with a Parasite-Bird Model

Mortality	Total Ex.	Parasite Ex.	Bird Ex.
$0.25 \rightarrow 1.0$	7.93	1.45	6.48
$1.00 \rightarrow 0.25$	8.49	1.58	6.91
0.625	8.31	1.44	6.87
0.25	7.85	1.95	5.90

of biological diversity is not latitude per se, but the environmental variables correlated with latitude. More than 25 different mechanisms have been suggested for generating latitudinal diversity gradients, but no consensus has been reached yet (Gaston, 2000). One factor proposed as very important for the latitudinal diversity gradients is the area of the climatic zones. Tropical landmasses have a larger climatically similar total surface area than landmasses at higher latitudes with similarly small temperature fluctuations (Rosenzweig, 1992). This may be related to higher levels of speciation and lower levels of extinction in the tropics (Rosenzweig, 1992; Gaston, 2000; Buzas et al., 2002). Most of the land surface of the earth was tropical or subtropical during the Tertiary, which could in part explain the greater diversity in the tropics today. The higher solar radiation in the tropics increases productivity, which in turn increases biological diversity. However, productivity can only explain why there is more total biomass in the tropics, not why this biomass should be allocated into more individuals, and these individuals into more species (Blackburn and Gaston, 1996). Body sizes and population densities are also typically lower in the tropics, implying a higher number of species, but the causes and the interactions among the many variable factors are complex and still uncertain (Blackburn and Gaston, 1996). Higher temperatures in the tropics may imply shorter generation times and greater mutation rates, thus accelerating speciation in the tropics (Rohde, 1992). Higher temperature means also that the biological processes are faster, and as the evolution resulting from biological processes in general increases the number of species (see Chapter 10), a higher temperature should imply a higher probability for a higher biodiversity. Speciation may also be accelerated by a higher habitat complexity in the tropics, although this does not seem to apply to freshwater ecosystems. The most likely explanation is a combination of various factors, and it is expected that different factors affect differently different groups of organisms, regions (e.g., northern vs. southern hemisphere), and ecosystems, yielding the variety of patterns that we observe. Anyhow, we can to a high extent apply system ecology and the presented principles to explain these observations about the latitudinal gradients in biodiversity. Several specific examples are given in Jørgensen et al. (2007).

7. MacArthur and Pianka (1966) first proposed an optimal foraging theory, arguing that because of the key importance of successful foraging to an individual's survival, it should be possible to predict foraging behavior by using theoretical considerations to determine how to maximize usable food intake. In their paper, a graphical model of animal feeding activities based on costs vs. profits was developed, implying that the strategy that would win would correspond to the strategy giving the highest survival, and therefore eco-exergy according to ELT. A forager's optimal diet was specified and some interesting predictions emerged. Prey abundance influenced the degree to which a consumer could afford to be selective, because it affected search time per item eaten. Diets should be broad when prey is scarce (long search time), but narrow if food is abundant (short search time), because a consumer can afford to bypass inferior prey only when there is a reasonably high probability of encountering a superior item in the time it would have taken to capture and handle the previous one. Also, larger patches

should be used in a more specialized way than smaller patches because travel time between patches (per item eaten) is lower. Three factors are important in this context:

a. How long a predator will forage in a specific area
b. Influence of prey density on the length of time a predator will forage in an area
c. Influence of prey variety on a predator's choice of acquired prey

These factors describe a predator's behavior as a function of its relationship with the prey it acquires. Fundamental conditions in these concepts influencing the predator-prey relationship are time foraging and prey availability. The observations are embodied in the optimal foraging theory. There are numerous examples that illustrate this theory. A typical example is given in Example 15.1, and more examples can be found in Jørgensen et al. (2007). One more clear example will, however, be given here. Kacelnik (1984) examined the optimal foraging theory by a clever experiment with starlings. He placed feeding tables at different distances from nests. He found that the starlings took larger loads of food, requiring more searching time, when the food source was more distant (requiring longer travel time). The starlings optimized the amount of food they could deliver to the nests per unit of time, which would yield the highest growth rate of the offspring. Another example was given as example 9 in Section 7.3.

8. The most successful organisms (ants, bacteria, and cockroaches, for instance) are the ones that utilize their free energy resources most efficiently. So, their success can be explained by their ability to maintain and even increase the eco-exergy.

9. If land is contaminated by heavy metals, a relatively high plant biomass will be maintained, which means a high eco-exergy, in spite of the contamination, because a selection of plants that will produce special proteins that can bind the metals will take place. Particularly, the grasses with these properties need not have very large genetic changes to produce the metal binding proteins. This adaptation is an example in accordance with ELT.

Example 15.1

Richardson and Verbeek (1986) have observed that crows in the Pacific Northwest often leave littleneck clams uneaten after locating them. The crows dig the clams from their burrows, but they often leave the smaller ones on the beach and only bother with the larger ones, which they drop on the rocks and eat. Their acceptance rate increases with prey size: they open and eat only about half of the 29 mm long clams they find, while consuming all clams in the 32–33 mm range. The most profitable clams were the largest, not because they broke more easily, but because they contained more free energy (chemical energy) than smaller clams. By considering the caloric benefits from clams of different sizes and the costs of searching for, digging up, opening, and feeding on clams, it is possible to construct a mathematical model based on the assumption that crows would select an optimal diet, which maximizes their caloric intake (see Figure 15.2).

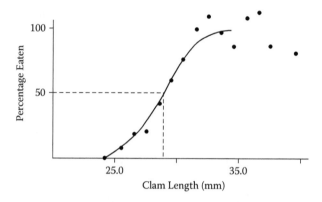

FIGURE 15.2   The graph shows the predicted percentages of clams that crows should select for consumption after locating, based on the assumption as function of the clam size. The birds attempt to maximize the rate of energy gain per unit of time spent foraging for clams. The dots represent the actual observations, and the graph the results of the developed model. (From Richardson, H., and Verbeek, N.A., *Ecology*, 67, 1219–1226, 1986.)

> Briefly explain that this example complies with the optimum foraging theory and ELT. Answer particularly the following question: Why is it not beneficial for the crows to eat the small clams?

Solution

The model considers the search costs required to find clams of different sizes. By investing more time in searching for bigger clams, the crows will be able to find bigger clams that give them chemical energy that more than compensates for the longer time. If the crows ate the small clams, they would use more energy to break them on the rocks and afterwards eat them than they would gain by the caloric content of the clam biomass. The optimum foraging strategy entails that the crows are able to adjust the prey size to what gives the most energy per unit of time, which the observations and the model confirm. Most energy per unit of time complies with ELT.

15.3 Application of Systems Ecology to Explain the Principles Applied in Ecological Engineering

Mitsch and Jørgensen (2004) have presented 19 principles that should be taken into account in the use of ecological engineering (eco-technology) or in the management of ecosystems. The 19 principles are strongly interrelated to the properties of ecosystems, because when we are applying engineering in ecosystems we have to consider their characteristic properties.

It is the idea that the 19 rules should be carefully followed in ecological engineering. They can be considered recommendations to environmental managers when they are applying ecological engineering, or they may be considered a checklist for the application of ecological engineering. These rules are rooted in the properties of the ecosystems, and in systems ecology. The explanation of the principles and their basis in systems ecology is given below.

Principle 1: The external factors (in ecology denoted the forcing functions) determine the reactions of ecosystems. Rooted in the openness of the ecosystem (Chapter 8). It is important when we discuss the possibilities to make man-controlled systems more sustainable that we realize that, by regulation of the forcing functions, we can change the reactions of the man-controlled systems.

Principle 2: Available storage of matter and energy is limited. Explained by the thermodynamic laws and the ability of the ecosystem to recycle (Chapter 2).

Principle 3: Ecosystems are open systems and have to be considered in all management plans. This is completely consistent with the openness of ecosystems in general.

Principle 4: Ecosystems are homeostatic systems with much feedback. Complies with the system properties of ecosystems, as presented in Chapter 14.

Principle 5: Homeostasis of ecosystems requires accordance between biological function and chemical composition of forcing functions. Rooted in the biochemistry that ecosystems have to follow (see Chapter 5).

Principle 6: Ecosystems recycle all matter and energy. Explained by the network properties of ecosystems (see Chapter 12).

Principle 7: Ecosystems are self-designing systems. An emergent holistic property of ecosystems (see Chapter 14).

Principle 8: Ecosystems have characteristic scales in space and time. Clearly in accordance with the hierarchical organization of ecosystems (see Chapter 9).

Principle 9: Ecosystems have high diversity and complexity. The biodiversity should be championed to maintain an ecosystem self-design capacity. Presented directly in Chapter 10. The high complexity is furthermore a holistic emerging property.

Principle 10: Ecotones, transition zones, are as important for ecosystems as membranes are for cells. This principle is consistent with the openness of ecosystems, which of course makes ecosystems vulnerable (see Chapter 8).

Principle 11: Due to their openness, ecosystems are coupled with other ecosystems. Also rooted in the openness (Chapter 8).

Principle 12: Ecosystems form networks. The components of ecosystems are connected.

Principle 13: Ecosystems have a history of development. The application of eco-technology should respect that. A part of the history is explained by the growth and development of ecosystems (Chapter 6). Another part is due to the change in the external factors, the openness (Chapter 8). The development of ecosystems has ensured the maintenance of the functions—primary production, secondary production, and so on—not certain species. The maintenance of the function can actually be considered a consequence of the high buffer capacity and resilience of ecosystems (see Chapter 11).

Principle 14: Ecosystems and their species are most vulnerable at their geographical edge. Due to the lower buffer capacity and resilience, when the conditions are unfavorable. This is touched upon in Chapter 11.

Principle 15: Ecosystems are hierarchical systems. This is directly covered in Chapter 9. This form of organization is important to consider in ecological management.

Principle 16: Ecosystems are pulsing systems. Due to the pulsing of the forcing functions and therefore to the openness.

Principle 17: Physical and biological processes are interactive. Rooted in the enormous variability of the forcing functions presented in Chapter 10.

Principle 18: Ecosystems are integrated systems with holistic properties. This is directly presented in Chapter 14 about holistic emerging properties.

Principle 19: Ecosystems store their information in the structure. Directly treated and discussed in Chapter 13.

15.4 Application of Systems Ecology to Assess Ecosystem Health

Ecosystems can be assessed by means of ecological indicators and by the calculation of the value of ecosystem services. The latter has been touched upon in Chapter 10, where the total production expressed as eco-exergy was calculated (see Table 10.4). As the value of 1 kWh is about 4 eurocents, or 5 dollar cents, it is possible to find the value of all possible ecosystem services by multiplying the annual production expressed in GJ/(ha year) by 10 euro or 12.5 dollars. The results are shown in Table 15.2 (see Jørgensen, 2010) together with the values of ecosystem services found by Costanza et al. (1997). They were calculating the values of the actual services that various ecosystems provided to the society. The value based on the total eco-exergy produced by the different ecosystems is of course higher, because it includes not only the ecosystem services that society is utilizing, but all possible—utilized or not utilized by society—ecosystem services. Notice that the ratio of the two values is about 10-30 (lakes, rivers, coastal zones, and wetlands)

TABLE 15.2 Approximate Value of Annual Ecosystem Services

Ecosystem	kEURO/ha year (based on eco-exergy)	$/ha year (according to Costanza et al., 1997)
Desert	20.7	?
Open sea	23.8	252
Coastal zones	48.3	4,052
Coral reefs, estuaries	960	14,460
Lakes, rivers	85	8,500
Coniferous forests	739	969
Deciduous forests	1,320	969
Temperate rain forests	1,580	?
Tropical rain forests	3,200	2,007
Tundra	72.8	?
Croplands	400	92
Grassland	175	232
Wetlands	450	14,785

TABLE 15.3 Classification of Ecological Indicators

Level	Example
Reductionistic (single) indicators	PCB, species present/absent, primary production
Semiholistic indicators	Odum's attributes
Holistic indicators	Biodiversity, ecological network, buffer capacity
Super holistic	Thermodynamic indicators as eco-exergy and emergy

for the ecosystems that society is using heavily, while the ratio is much higher for the ecosystems that are not utilized very much by society, or only used for biomass production (for instance, tropical rain forests and agricultural ecosystems).

Ecological indicators may be classified into four groups (see Table 15.3). The selections of the reductionistic indicators are often obvious because they are based on the pollution problem directly, for instance, a toxic substance, or eutrophication, for instance, primary production or concentration of phytoplankton or transparency.

The other classes of indicators are, however, based on system ecology.

1. Odum's attributes (see Table 2.5), for instance, express in more detail the three growth forms, as discussed in Section 2.6. Several of the attributes are also directly used as holistic indicators; see point 2.
2. The holistic indicators biodiversity, ecological networks, and buffer capacity are directly diverted from ecosystem properties that are crucial for the growth and development of ecosystems. All three are covered in Chapters 10–12.
3. The thermodynamic indicators are rooted in the importance of thermodynamics for the understanding of ecosystems, including the application of the thermodynamic laws and ELT on ecosystems. Eco-exergy and emergy are presented in Chapters 3 and 4 and again touched on in Chapters 6 and 7, where ELT is presented in detail. Emergy is used as an indicator to give the costs of the flows (power) expressed in solar energy equivalents or the capital that nature has invested expressed in solar energy. The units used for these two applications of emergy are sej/year and sej. Eco-exergy, on the other hand, gives the production of eco-exergy per unit of time or the power expressed in kJ/year of the capital of eco-exergy or work capacity that the ecosystem has invested as biomass, information, and structure of the network, expressed in kJ. Eco-exergy is often also expressed as eco-exergy density, which is the eco-exergy per unit of area or per unit of volume.

Bastianoni and Marchettini (1997) have proposed to use the ratio eco-exergy/emergy power (emergy flows) as indicators. The ratio would indicate how much structure in J can be maintained far from thermodynamic equilibrium per external input expressed as sej/ unit of time. Emergy is used in this context because solar radiation is the driving force of all the energy flows. Table 15.4 shows the results obtained by Bastianoni and Marchettini.

As seen, the natural systems show very similar figures, while the man-controlled systems have a much smaller ratio. Natural systems where selection has worked relatively undisturbed for a long time have much higher thermodynamic efficiency than man-controlled systems; see also the discussion in Jørgensen et al. (2007).

TABLE 15.4 Empower Density, Eco-Exergy Density, and Eco-Exergy to Empower Ratio for 8 Different Ecosystems

	Control Pond	Waste Pond	Caprolace Lagoon	Trasimeno Lake	Venice Lagoon	Figheri Basin	Iberá Lagoon	Galarza Lagoon
Empower density (sej/year·L)	$20.1{\cdot}10^8$	$31.6{\cdot}10^8$	$0.9{\cdot}10^8$	$0.3{\cdot}10^8$	$1.4{\cdot}10^9$	$12.2{\cdot}10^8$	$1.0{\cdot}10^8$	$1.1{\cdot}10^8$
Eco-exergy density (J/L)	$1.6{\cdot}10^4$	$0.6{\cdot}10^4$	$4.1{\cdot}10^4$	$1.0{\cdot}10^4$	$5.5{\cdot}10^4$	$71.2{\cdot}10^4$	$7.3{\cdot}10^4$	$5.5{\cdot}10^4$
Eco-exergy/empower (10^{-5}J·year/sej)	0.8	0.2	44.3	30.6	39.1	58.5	73	50.0

Example 15.2

Coastal lagoons are subjected to strong anthropogenic pressure. This is partly due to freshwater input rich in organic and mineral nutrients derived from urban, agricultural, or industrial effluent and domestic sewage, but also due to the intensive shellfish farming. The Sacca di Goro is a shallow water embayment of the Po Delta. The surface area is 26 km² and the total water volume is approximately 40×10^6 m³. The catchment basin is heavily exploited by agriculture, while the lagoon is one of the most important clam (*Tapes philippinarum*) aquaculture systems in Italy. The combination of all these anthropogenic pressures calls for an integrated ecological and environmental management that considers all different aspects—lagoon fluid dynamics, ecology, nutrient cycles, river runoff influence, shellfish farming, macroalgal blooms, and sediments, as well as the socioeconomical implication of different possible management strategies. All these factors are responsible for important disruptions in ecosystem functioning characterized by eutrophic and dystrophic conditions in summer (Viaroli et al., 2001), algal blooms, oxygen depletion, and sulfide production (Chapelle et al., 2000). Water quality is the major problem. From 1987 to 1992 the Sacca di Goro experienced an abnormal proliferation of macroalga *Ulva* spp. The massive presence of this macroalga heavily affected the lagoon ecosystem, and it was necessary to remove the *Ulva* biomass mechanically.

To carry out an integrated environmental management, a biogeochemical model, partially calibrated and validated with field data from 1989 to 1998, was developed (Zaldívar et al., 2003). To analyze its results, it is necessary to utilize ecological indicators, using not only indicators based on particular species or components (macrophytes or zooplankton), but also indicators able to include structural, functional, and system level aspects. Eco-exergy and specific eco-exergy (eco-exergy/biomass) were used to assess the ecosystem health of this coastal lagoon. Three scenarios were analyzed (for a system with clam production and eutrophication by *Ulva*) using a lagoon model: (1) present situation, (2) optimal strategy based on cost-benefit for removal of *Ulva*, and (3) a significant nutrient loading reduction from watershed. The cost-benefit model gave the direct cost of *Ulva* harvesting, including vessel cost and damage to shellfish production and the subsequent mortality increase in the clam population. The total benefit obtained from simulating the biomass

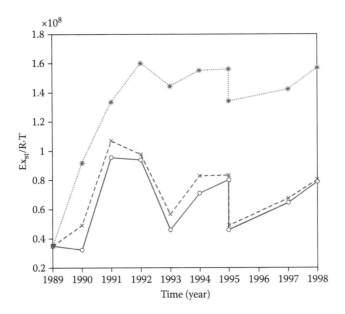

FIGURE 15.3 (See color insert.) Eco-exergy mean annual values from 1989 to 1998: present scenario (continuous line), removal of *Ulva*, optimal strategy from cost-benefit point of view (dotted line), and nutrients load reduction from watershed (dashed line).

increase was evaluated using the averaged prices for clams in the northern Adriatic; therefore, an increase in clam biomass harvested from the lagoon will result in an increase of benefit. The Sacca di Goro model has several state variables for which the eco-exergy was computed: Figures 15.3 and 15.4 present the evolution of exergy and specific exergy for the two proposed scenarios: *Ulva* removal and nutrient load reduction, in comparison with the "no changes" alternative.

Conclusions of these investigations included the model results for the three scenarios by the application of ELT and comment on the applicability of eco-exergy and specific eco-exergy as indicators. This case study was presented by Zaldivar et al. in *Handbook of Ecological Indicators for Assessment of Ecosystem Health*, by Jørgensen, Costanza, and Xu, CRC Press, Boca Raton, FL, 2005, pp. 163–184.

Solution

The results show that the cost-benefit optimal solution for removal of *Ulva* has the highest eco-exergy and specific eco-exergy, followed by a significant removal of nutrients from the watershed. In the case for removal of *Ulva* specific exergy continues to increase as the number of vessels operating in the lagoon increases. The present situation (no changes) had the lowest eco-exergy and specific eco-exergy. The result shows that it is a good sustainability policy to take care of natural resources, in this case the clams. Eco-exergy expresses the system biomass and genetic information embedded in that biomass, while

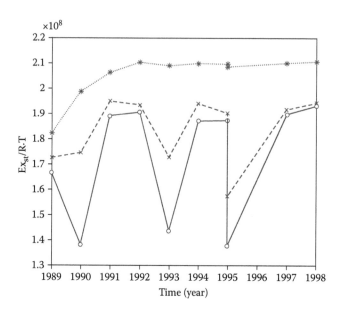

FIGURE 15.4 (See color insert.) Specific eco-exergy mean annual values from 1989 to 1998: present scenario (continuous line), removal of *Ulva*, optimal strategy from cost-benefit point of view (dotted line), and nutrients load reduction from watershed (dashed line).

specific eco-exergy tells us how rich in information the system is. The results are consistent with ELT. The applied indicators broadly encompass ecosystem characteristics, and it has been shown that they are correlated with several important parameters, such as respiration, biomass, etc. However, it has been pointed out (Jørgensen, 2000) that eco-exergy is not directly related to biodiversity, and a very eutrophic system often has a low biodiversity but high eco-exergy due to the high biomass. The two indicators were beneficially applied in this case for a sustainable ecological and environmental management.

15.5 Summary of Important Points in Chapter 15

1. Systems ecology has found a wide application in integrated ecological and environmental management. It is working hand in hand with the new emergent ecological subdisciplines, which are indispensable tools for the integrated holistic management: ecological modeling, ecological engineering, assessment of ecosystem health, and ecosystem services.
2. The up-to-date integrated ecological and environmental management consists of seven steps:
   a. Define the problem.
   b. Determine the ecosystems involved.

c. Find and quantify all the sources to the problem.

d. Set up a diagnosis to understand the relation between the problem and the sources.

e. Determine all the tools we need to implement to solve the problem.

f. Implement the selected solutions.

g. Follow the recovery process.

3. For the diagnosis (point 4) three tools are used: ecological modeling, ecological indicators, and ecological services.

4. The toolboxes applied to solve the environmental problems are environmental technology, ecological engineering, environmental legislation, and cleaner technology. The latter three are relatively new toolboxes.

5. Ecological engineering follows 19 principles that are also used as a checklist for the sustainability of an ecological engineering project. These 19 principles are all directly rooted in systems ecology.

6. Ecological indicators can be divided into four classes: reductionistic indicators, semiholistic indicators, holistic indicators, and super holistic or thermodynamic indicators. The three latter classes are all rooted in systems ecology.

7. The indicator eco-exergy density/emergy power shows very clearly a significant difference between natural systems and man-controlled systems. The indicator expresses the thermodynamic efficiency of ecosystems, and it is considerably higher for natural ecosystems.

Exercises/Problems

1. How would a high concentration in a river of a toxic substance effect eco-exergy and specific eco-exergy? It is presumed that they are applied as indicators.

2. The physical variables with a river system present a continuous gradient of physical conditions. This gradient should, according to the river continuum concept (RCC), elicit a series of responses resulting in a continuum of biotic adjustments and consistent patterns of loading, transport, utilization, and storage of organic matter along the length of the river. The biological systems of the river move toward a balance between a tendency for efficient use of energy inputs through resource partitioning (particularly food) and an opposing tendency for a uniform rate of energy processing throughout the year. RCC assumes that natural streams process strategies involving minimum energy loss. Comment on the RCC in the light of ELT.

3. Is Liebig's law (see Chapter 2) in accordance with ELT?

4. Indicate the difference between a natural wetland and an agricultural field. Propose three indicators and indicate the difference between these indicators for the two ecosystems.

5. Most environmental management cases require more than one or two indicators. Why?

6. Ascendency is also used as an indicator. In which cases is it probably beneficial to use ascendency as an indicator for an ecosystem?

References

Ahl, T., and Weiderholm, T. 1977. *Svenska vattenkvalitetskriterier. Eurofierande ämnen.* SNVPM (Sweden). Report 918.

Allen, E. et al. 2011. More diverse plant communities have a higher functioning over time due to turnover in complementary dominant species. *Proceedings of the National Academy of Sciences.* 108, 17034–17039.

Allen, T.F.H., and Starr, T.B. 1982. *Hierarchy, perspectives for ecological complexity.* The University of Chicago Press, Chicago and London.

Andresen, I. 1983. Optimization of exergy. Doctor of science thesis at Copenhagen University.

Aoki, I. 1987. Entropy balance in Lake Biva. *Ecol. Model.* 37, 235–248.

Aoki, I. 1988. Entropy laws in ecological networks at steady state. *Ecol. Model.* 42, 289–303.

Aoki, I. 1989. Ecological study of lakes from an entropy viewpoint—Lake Mendota. *Ecol. Model.* 49, 81–87.

Bak, P. 1996. *How nature works.* Springer Verlag, New York.

Bastianoni, S., and Marchettini, N. 1997. Emergy/exergy ratio as a measure of the level of organization of systems. *Ecol. Model.* 99, 33–40.

Benndorf, J. 1987. Food-web manipulation without nutrient control: a useful strategy in lake restoration? *Schweiz Z. Hydrol.* 49, 237–248.

Benndorf, J. 1990. Conditions for effective biomanipulation. Conclusions derived from whole-lake experiments in Europe. *Hydrobiologia*, 200/201, 187–203.

Blackburn, T.M., and Gaston, K.J. 1996. A sideways look at patterns in species richness, or why there are so few species outside the tropics. *Biodiversity Lett.* 3, 44–53.

Boltzmann, L. 1905. The second law of thermodynamics (Populare schriften. Essay no. 3 (Address to Imperial Academy of Science in 1886)). Reprinted in English in *Theoretical physics and philosophical problems, selected writings of L. Boltzmann.* D. Riedel, Dordrecht.

Bonner, J.T., 1965. Size and cycle. *An essay on the structure of biology.* Princeton University Press..

Brillouin, L. 1962. *Science and information theory.* 2nd ed. Academic Press, New York.

Brown, J.H. 1995. *Macroecology.* University of Chicago Press, Chicago.

Brown, J.H., Marquet, P.A., and Taper, M.L. 1993. Evolution of body size: consequences of an energetic definition of fitness. *Am. Nat.* 142, 573–584.

Brown, M.T., and McClanahan, T.R. 1992. *Energy systems. Overview of Thailand.* Center for Wetlands, University of Florida, Gainesville, Florida.

Buzas, M.A., Collins, L.S., and Culver, S.J. 2002. Latitudinal difference in biodiversity caused by higher tropical rate of increase. *Proc. Natl. Acad. Sci.* 99, 7841–7843.

Calaco, A., et al. 2007. Polar fatty acids as indicator of trophic associations in a deep-sea vent system community. *Marine Ecol.* 28, 15–24.

Capek, V., and Sheehan, D.P. 2005. *Challenges to the second law of thermodynamics. Theory and experiment. Fundamental theories of physics.* Vol. 146. Springer, New York.

Cerbe, G.., and Hoffmann H.J. 1996. *Einführung in the thermodynamik.* (11th ed.) Carl Hanser Verlag München/Wien.

Chapelle, A., Ménesguen, A., Deslous-Paoli, J.M., Souchu, P., Mazouni, N., Vaquer, A., and Millet, B. 2000. Modelling nitrogen, primary production and oxygen in a Mediterranean lagoon. Impact of oysters farming and inputs from the watershed. *Ecol. Model.* 127, 161–181.

Childress, J.J., Felbeck, H., and Somero, G.N. 1987. Symbiosis in the deep sea. *Scientific American*, 256, 39–48.

Christensen, V., and Pauly, D., 1992. Ecopath II: A software for balancing steady-state ecosystem models and calculating network characteristics. *Ecol. Model.* 61, 169–185.

Commoner, B. 1971. *The closing circle.* Alfred A. Knopf. New York.

Cook, L.M. 2000. Changing views on melanic moths. *Biol. J. Linn. Soc.* 69, 431–444.

Cornell, J.H. 1978. Diversity in tropical rainforests and coral reefs. *Science* 99, 1302–1310.

Costanza, R., and Sklar, F.H. 1985. Articulation, accuracy and effectiveness of mathematical models: a review of freshwater wetland applications. *Ecol. Model.* 27, 45–69.

Costanza, R., Norton, B.G., and Haskell, B.D. 1992. *Ecosystem health, new goals for environmental management.* Island Press, Washington, D.C.

Costanza, R., et al. 1997. The value of the world's ecosystem services and natural capital. *Nature* 387, 252–260.

Covich, A.P. 2010. Winning the biodiversity arms race among freshwater gastropods: competition and coexistence through shell variability and predator avoidance. *Hydrobiologia*, 653, 191–215.

Coyne, J.A. 2002. Evolution under pressure. *Nature*, 418, 19–20.

Debeljak, M. 2002. Application of exergy degradation and exergy storage as indicator for the development of managed and virgin forest. PhD thesis at Ljubliana University.

de Bernardi, R. 1989. Biomanipulation of aquatic food chains to improve water quality in eutrophic lakes. In *Ecological assessment of environmental degradation, pollution and recovery*, ed. O. Ravera, 195–215. Elsevier Sci. Publ., Amsterdam.

de Bernardi, R., and Giussani, G. 1995. Biomanipulation: bases for a top-down control. In *Guidelines of lake management: biomanipulation in lakes and reservoirs*, ed. R. De Bernardi and G. Giussani, 1–14. Vol. 7. ILEC and UNEP, Kusatsu, Japan.

De Duve, C. 2002. *Life evolving.* Oxford University Press, Oxford.

Devillers, J., (ed.) 2009. *Ecotoxicological modeling,* Springer Verlag. Heidelberg, Berlin.

Ebeling, W., Engel, A., and Feistel, R. 1990. *Physik der evolutionsprozesse.* Akademie Verlag Berlin.

Eigen M., with R. Winkler-Oswatitsch. 1992. *Steps toward life: a perspective on evolution.* Oxford University Press, Oxford.

Fath, B.D. 2004. Editorial: control of distributed systems and environmental applications. *Ecol. Model.* 179, 151–152.

Fath, B.D., and Patten, B.C. 1999. Review of the foundations of network environ analysis. *Ecosystems* 2, 167–179.

Fenchel, T. 1974. Intrinsic rate of natural increase: the relationship with body size. *Oecologia*, 14, 317–326.

Fisher, J.C., and R.A. Hinde. 1949. The opening of milk bottles by birds. *British Birds.* 42, 347–357.

Futuyma, D.J. 1986. *Evolutionary biology.* 2nd ed. Sinauer Associates, Sunderland, MA.

Gaston, K.J. 2000. Global patterns in biodiversity. *Nature* 405, 220–227.

Gause, G.F. 1934. *The struggle for existence.* Williams and Wilkins, Baltimore.

Geigy. 1990. *Scientific Tables.* (9th ed.). Edited by K. Diem and C. Lentner. J.R. Geigy. Basel, Switzerland. 840 pp.

Gibbons, D.W., Reid, J.B., and Chapman, R.A. 1993. *The new atlas of breeding birds in Britain and Ireland: 1988–1991.* Poyser, London.

Gilliland, M.W. 1982. *Embodied energy studies of metal and fuel minerals.* Report to National Science Foundation.

Giussani, G., and Galanti, G. 1995. Case study: Lake Candia (Northern Italy). In *Guidelines of lake management: biomanipulation in lakes and reservoirs*, ed. R. De Bernardi and G. Giussani, 135–146. Vol. 7. ILEC, Kusatsu, Japan.

Givnish, T.J., and Vermelj, G.J. 1976. Sizes and shapes of liana leaves. *Am. Nat.* 110, 743–778.

Glansdorff, P., and Prigogine, I. 1971. *Thermodynamics of structure, stability and fluctuations.* Wiley Interscience Publishers, New York.

Grant, B.S. 2002. Sour grapes of wrath. *Science* 297, 940–941.

Grant, P.R. 1986. *Ecology and evolution of Darwin's finches.* Princeton University Press, Princeton, NJ. Reprinted in 1999.

Hall, C.A.S. 1995. *Maximum power—the ideas and applications of H.T. Odum.* University Press of Colorado, Niwot.

Hannon, B. 1973. The structure of ecosystems. *J. Theor. Biol.* 41, 534–546.

Hannon, B. 1979. Total energy cost in ecosystems. *J. Theor. Biol.* 80, 271–293.

Hannon, B. 1982. Energy discounting. In *Energetics and systems*, ed. W. Mitsch, R. Ragade, R. Bosserman, and J. Dillon, 73–100. Ann Arbor Science Publishers, Ann Arbor, MI.

Hassell, M.P., Lawton, J.H., and May, R.M. 1976. Pattern of dynamical behaviour in single species populations. *J. Anim. Ecol.* 45, 471–486.

Hastie, N. 2001. Perspective. In *Essays in biochemistry*, ed. K.E. Chapman and S.J. Higgins, 121–128. Portland Press.

Haugaard Nielsen, R. 2001. Hunting the proteins [In Danish Jagten på proteinerne]. *Ingeniøren* 16, 20. April 20–21.

Havens, K.E. 1999. Structural and functional responses of a freshwater community to acute copper stress. *Environ. Poll.* 86, 259–266.

Hawking, S. 2001. *The universe in a nutshell.* The Book Laboratory Bantam, New York.

Herendeen, R. 1981. Energy intensity in ecological economic systems. *J. Theor. Biol.* 91, 607–620.

Heslop-Harrison, Y. 1978. Carnivorous plants. *Scientific American*, February.

Higashi, M., Patten, B.C., and Burns, T.P. 1989. Dominance of indirect causality in ecosystems. *Am. Nat.* 133, 288–302.

Hirschfelder, J.O., Curtiss, C.F., and Bird, R.B. 1954. *Molecular theory of gases and liquids.* John Wiley & Sons, New York.

Holling, C.S. 1986. The resilience of terrestrial ecosystems: local surprise and global change. In *Sustainable development of the biosphere*, ed. W.C. Clark and R.E. Munn, 292–317. Cambridge University Press, Cambridge.

Hosper, S.H. 1989. Biomanipulation, new perspective for restoring shallow, eutrophic lakes in the Netherlands. *Hydrobiol. Bull.* 73, 11–18.

Hutchinson, G.E. 1948. Circular causal systems in ecology. *Ann. NY Acad. Sci.* 50, 221–246.

Hutchinson, G.E. 1957. Concluding remarks. *Cold Spring Harbor Symp. Quant. Biol.* 22, 415–427.

Hutchinson, G.E. 1965. The niche: an abstractly inhabited hypervolume. In *The ecological theatre and the evolutionary play*, ed. G.E. Hutchinson, 26–78. Yale University Press, New Haven, CT.

Jablonka, E., and Lamb, M.J. 2006. *Evolution in four dimensions.* MIT Press, Boston.

Jensen Peter, K.A. 2005. *Human origin and evolution* [Menneskets Oprindelse og udvikling]. 3rd ed. Gyldendal, Copenhagen.

Jeppesen, E.J., et al. 1990. Fish manipulation as a lake restoration tool in shallow, eutrophic temperate lakes. Cross-analysis of three Danish case studies. *Hydrobiologia* 200/201, 205–218.

Jørgensen, S.E. 1976a. A eutrophication model for a lake. *J. Ecol. Model.* 2, 147–165.

Jørgensen, S.E. 1976b. An ecological model for heavy metal contamination of crops and ground water. *Ecol. Model.* 2, 59–67.

Jørgensen, S.E. 1982. A holistic approach to ecological modelling by application of thermodynamics. In *Systems and energy*, ed. W. Mitsch et al. Ann Arbor Science. Ann Arbor, MI.

Jørgensen, S.E. 1984. Parameter estimation in toxic substance models. *Ecol. Model.* 22, 1–12.

Jørgensen, S.E. 1988. Use of models as experimental tools to show that structural changes are accompanied by increased exergy. *Ecol. Model.* 41, 117–126.

Jørgensen, S.E. 1990. Ecosystem theory, ecological buffer capacity, uncertainty and complexity. *Ecol. Model.* 52, 125–133.

Jørgensen, S.E. 1992a. Parameters, ecological constraints and exergy. *Ecol. Model.* 62, 163–170.

Jørgensen, S.E. 1992b. Development of models able to account for changes in species composition. *Ecol. Model.* 62, 195–208.

Jørgensen, S.E. 1994a. Review and comparison of goal functions in system ecology. *Vie Milieu* 44, 11–20.

Jørgensen, S.E. 1994b. Fundamentals of ecological modeling. 2nd ed., Developments in Environmental Modelling 19. Elsevier, Amsterdam.

Jørgensen, S.E. 1995a. Exergy and ecological buffer capacities as measures of the ecosystem health. *Ecosyst. Health* 1, 150–160.

Jørgensen, S.E. 1995b. The growth rate of zooplankton at the edge of chaos: ecological models. *J. Theor. Biol.* 175, 13–21.

Jørgensen, S.E. 1997. Thermodynamik offener Systeme. In *Grundlagen der Ökosystemtheorie III-1.6*, pp. 1–21. Springer, Heidelberg.

Jørgensen, S.E. 2000. *The principles of pollution abatement*. Elsevier, Amsterdam.

Jørgensen, S.E. 2002. *Integration of ecosystem theories: a pattern*. Kluwer, Dordrecht.

Jørgensen, S.E. 2006. *Eco-exergy as sustainability*. WIT, Southampton, UK.

Jørgensen, S.E. 2007. Description of aquatic ecosystem's development by eco-exergy and exergy destruction. *Ecol. Model.* 204, 22–28.

Jørgensen, S.E. 2008. *Evolutionary essays*. Elsevier, Amsterdam.

Jørgensen, S.E. 2009. The application of structurally dynamic models in ecology and eco-toxicology. In *Ecotoxicological modeling*, ed. J. Devillers, 377–394. Springer Verlag.

Jørgensen, S.E. 2010. Ecosystem services, sustainability and thermodynamic indicators. *Ecol. Complexity* 7, 311–313.

Jørgensen, S.E., and Bendoricchio, G. 2001. *Fundamentals of ecological modelling*. 3rd ed. Elsevier, Amsterdam.

Jørgensen, S.E., and de Bernardi, R. 1997. The application of a model with dynamic structure to simulate the effect of mass fish mortality on zooplankton structure in Lago di Annone. *Hydrobiologia* 356, 87–96.

Jørgensen, S.E., and de Bernardi, R. 1998. The use of structural dynamic models to explain successes and failures of biomanipulation. *Hydrobiologia* 359, 1–12.

Jørgensen, S.E., and Fath, B.D. 2004. Modelling the selective adaptation of Darwin's finches. *Ecol. Model.* 176, 409–418.

Jørgensen, S.E., and Fath, B. 2006. Examination of ecological networks. *Ecol. Model.* 196, 283–288.

Jørgensen, S.E., and Fath, B. 2011. *Fundamentals of ecological modelling*. 4th ed. Elsevier, Amsterdam.

Jørgensen, S.E., Fath, B., Bastiononi, S., Marques, J.C., Mueller, F., Nielsen, S.N., Patten, B.C., Tiezzi, E., and Ulanovicz, R.E. 2007. *A new ecology*. Elsevier, Amsterdam.

Jørgensen, S.E., Ladegaard, N., Debeljak, M., and Marques, J.C. 2005. Calculations of exergy for organisms. *Ecol. Model.* 185, 165–176.

Jørgensen, S.E., Ludovisi, A., and Nielsen, S.N. 2010. The free energy and information embodied in the amino acid chains of organisms. *Ecol. Model.* 221, 2388–2392.

Jørgensen, S.E., and Mejer, H.F. 1977. Ecological buffer capacity. *Ecol. Model.* 3, 39–61.

Jørgensen, S.E., and Mejer, H.F. 1979. A holistic approach to ecological modelling. *Ecol. Model.* 7, 169–189.

Jørgensen, S.E., Mejer, H., and Friis, M. 1978. Examination of a lake. *J. Ecol. Model.* 4, 253–279.

Jørgensen, S.E., and Nielsen, S.N. In press. Tool boxes for an integrated ecological and environmental management. In *Ecological indicators*.

Jørgensen, S.E., Nielsen, S.N., and Jørgensen, L.A. 1991. *Handbook of ecological parameters and ecotoxicology*. Elsevier, Amsterdam. Published as CD under the name ECOTOX, with L.A. Jørgensen as first editor, in year 2000.

Jørgensen, S.E., Nielsen, S.N., and Mejer, H. 1995. Emergy, environ, exergy and ecological modelling. *Ecol. Model.* 77, 99–109.

Jørgensen, S.E., and Padisák, J. 1996. Does the intermediate disturbance hypothesis comply with thermodynamics? *Hydrobiologia* 323, 9–21.

Jørgensen, S.E., Patten, B.C., and Straskraba, M. 1999. Ecosystem emerging. 3. Openness. *Ecol. Model.* 117, 41–64.

Jørgensen, S.E., Patten, B.C., and Straškraba, M. 2000. Ecosystems emerging. 4. Growth. *Ecol. Model.* 126, 249–284.

Jørgensen, S.E., and Svirezhev, Y. 2004. *Toward a thermodynamic theory for ecological systems.* Elsevier, Amsterdam.

Jørgensen, S.E., and Ulanowicz, R. 2009. Network calculations and ascendency based on eco-exergy. *Ecol. Model.* 220, 1893–1896.

Kacelnik, A. 1984. Central place foraging in starlings. I. Patch residence time. *J. Animal Ecol.* 53, 283–299.

Kallqvist, T., and Meadows, B.S. 1978. The toxic effects of copper on algae and rotifers from a soda lake. *Wat. Res.* 12, 771–775.

Kauffman, S.A. 1993. *The origins of order.* Oxford University Press, Oxford.

Kauffman, S.A. 1995. *At home in the universe: the search for the laws of self organization and complexity.* Oxford University Press, New York.

Kay, J., and Schneider, E.D. 1992. Thermodynamics and measures of ecological integrity. In *Proceedings of "Ecological Indicators,"* 159–182. Elsevier, Amsterdam.

Kay, J.J. 1984. Self organization in living systems. PhD thesis, Systems Design Engineering, University of Waterloo, Ontario.

Kazanci, C. 2007. EcoNet: a new software for ecological modeling, simulation and network analysis. *Ecol. Model.* 208, 3–8.

Kazanci, C. 2009. Network calculations II: a user's manual for EcoNet. In *Handbook of ecological modelling and informatics*, ed. S.E. Jørgensen, T.-S. Chon, and F. Recknagel, chap. 18. WIT Press, Southampton, UK.

Kettlewell, H.B.D. 1965. Insect survival and selection for pattern. *Science* 148, 1290–1296.

Klipp, E., Herwig, R., Kowald, A., Wierling, C., and Lehrach, H. 2005. *Systems biology in practice. Concepts, implementation and application.* Wiley-VCH Verlag GmbH, Weinheim, Germany.

Koschel, R., Kasprzak, Krienitz, L., and Ronneberger, D. 1993. Long term effects of reduced nutrient loading and food-web manipulation on plankton in a stratified Baltic hard water lake. *Verh. Int. Ver. Limnol.* 25, 647–651.

Laszlo, E. 2003. *The connectivity hypothesis.* State University of New York Press, New York.

Leveque, C., and Mounolou, J.-C. 2003. *Biodiversity.* Wiley, New York.

Li, W.-H., and Grauer, D. 1991. *Fundamentals of molecular evolution.* Sinauer, Sunderland, MA.

Lorentz, E.H. 1955. Available potential energy and the maintenance of the general circulation. *Tellus* VII, 157–167.

Lorentz, E.H. 1963. Deterministic nonperiodic flow. *J. Atmosph. Sci.* 20, 130–141.

Lotka, A.J. 1956. *Elements of mathematical biology.* Dover. New York.

Ludovisi, A., and Jørgensen, S.E. 2009. Comparison of exergy found by a classical thermodynamic approach and by the use of the information stored in the genome. *Ecol. Model.* 220, 1897–1903.

MacArthur, R.H., and Pianka, E.R. 1966. On optimal use of a patchy environment. *Am. Nat.* 100, 603–609.

Madsen, J. 2006. *Livets Udvikling A-Å* [Development of the life from A to Z]. Gyldendal, Copenhagen.

Majerus, M.E.N. 1998. *Melanism. Evolution in action.* Oxford University Press, Oxford.

Marchi, M., Jørgensen, S.E., Bécares, E., Corsi, I., Marchettini, N., and Bastiononi, S. 2011. Dynamic model of Lake Chozas (León, NW Spain): decrease in eco-exergy from clear to turbid phase due to introduction of exotic crayfish. *Ecol. Model.* 222, 3002-3010.

May, R.M. 1973. *Stability and complexity in model ecosystems.* Princeton University Press, Princeton, NJ.

Mayr, E. 2001. *What evolution is.* Basic Books, New York.

Mitsch, W.J., and Jørgensen, S.E. 2004. *Ecological engineering and ecosystem restoration.* John Wiley, New York.

Monod, J. 1972. *Chance and necessity.* Random House, New York.

Morowitz, H.J. 1968. *Energy flow in biology. Biological organisation as a problem in thermal physics.* Academic Press, New York.

Nielsen, S.N. 1992a. Application of maximum exergy in structural dynamic models. PhD thesis, National Environmental Research Institute, Denmark.

Nielsen, S.N. 1992b. Strategies for structural-dynamical modelling. *Ecol. Model.* 63, 91–102.

Nielsen, S.N. 2007. What has modern ecosystem theory to offer to cleaner production, industrial ecology and society? The views of an ecologist. *J. Cleaner Prod.* 15, 1639–1653.

Nielsen, S.N., and Mueller, F. 2009. Understanding the functional principles of nature— proposing another type of ecosystem services. *Ecol. Model.* 220, 1913–1925.

Nielsen, S.N., and Müller, F. 2000. Emergent properties of ecosystems. In *Handbook of ecosystem theories and management*, ed. S.E. Jørgensen and F. Müller, 195–216. Lewis Publishers, Boca Raton, FL.

Odum, E.P. 1959. *Fundamentals of ecology.* 2nd ed. W.B. Saunders, Philadelphia, PA.

Odum, E.P. 1969. The strategy of ecosystem development. *Science* 164, 262–270.

Odum, E.P. 1971. *Fundamentals of ecology.* 3rd ed. W.B. Saunders Co., Philadelphia, PA.

Odum, H.T. 1971. *Environment, power, and society.* Wiley Interscience Publishers, New York.

Odum, H.T. 1983. *System ecology.* Wiley Interscience Publishers, New York.

Odum, H.T. 1988. Self-organization, transformity, and information. *Science* 242, 1132–1139.

Odum, H.T. 1989. Experimental study of self-organization in estuarine ponds. In *Ecological engineering: an introduction to eco-technology*, ed. W.J. Mitsch, and S.E. Jørgensen, 291–340. Wiley, New York.

Odum, H.T. 1996. *Environmental accounting: emergy and environmental decision making.* John Wiley & Sons, New York.

Odum, H.T., and Pinkerton, R.C. 1955. Time's speed regulator: the optimum efficiency for maximum power output in physical and biological systems. *Am. Sci.* 43, 331–343.

O'Neill, R.V., DeAngelis, D.L., Waide, J.B., and Allen, T.F.H. 1986. A hierarchical concept of ecosystems. Princeton University Press, Princeton, NJ.

Onsager, L. 1931. Reciprocal relations in irreversible processes, I. *Phys. Rev.* 37, 405–426.

Ostwald, W. 1931. Gedanken zur Biosphäre. Wiederabdruck. BSB B.G. Teubner Verlagsgesellschaft, Leipzig.

Patten, B.C. 1978. Systems approach to the concept of environment. *Ohio J. Sci.* 78, 206–222.

Patten, B.C. 1981. Environs: the super-niches of ecosystems. *Am. Zool.* 21, 845–852.

Patten, B.C. 1982. Environs: relativistic elementary particles or ecology. *Am. Nat.* 119, 179–219.

Patten, B.C. 1985. Energy cycling in the ecosystem. *Ecol. Model.* 28, 1–71.

Patten, B.C. 1991. Network ecology: indirect determination of the life–environment relationship in ecosystems. In *Theoretical ecosystem ecology: the network perspective*, ed. M. Higashi and T.P. Burns, 288–351. Cambridge University Press, London.

Patten, B.C. 1992. Energy, emergy and environs. *Ecol. Model.* 62, 29–69.

Patten, B.C., and Auble, G.T. 1981. System theory of the ecological niche. *Am. Nat.* 117, 893–922.

Patten, B.C., Straskraba, M., and Jørgensen, S.E. 1997. Ecosystem emerging 1: conservation. *Ecol. Model.* 96, 221–284.

Peters, R.H. 1983. *The ecological implications of body size.* Cambridge University Press, Cambridge.

Picket, S.T.A., and White, P.S. 1985. Natural disturbance and patch dynamics: an introduction. In *The ecology of disturbance and patch dynamics*, ed. S.T.A. Picket and P.S. White, 3–13. Academic Press, New York.

Prigogine, I. 1980. *From being to becoming: time and complexity in the physical sciences.* Freeman, San Francisco, CA.

Richardson, H., and Verbeek, N.A. 1986. Diet selection and optimization by north-western crows feeding in Japanese littleneck clams. *Ecology*, 67, 1219–1226.

Ricklefs, R.E. 2000. *The economy of nature.* 5th ed. W.H. Freeman and Company, New York.

Rohde, K. 1992. Latitudinal gradients in species diversity: the search for the primary cause. *Oikos* 65, 514–527.

Rosenzweig, M.L. 1992. Species diversity gradients: we know more and less than we thought. *J. Mammal.* 73, 715–730.

Rosenzweig, M.L. 1996. *Species diversity in space and time.* Cambridge University Press, Cambridge.

Russel, L.D., and Adebiyi, G.A. 1993. *Classical thermodynamics.* Saunders College Publishing, Harcourt Brace Jovanovich College Publishers, Fort Worth, TX.

Rutledge, R.W., Basorre, B.L., and Mulholland, R.J. 1976. Ecological stability: an information theory viewpoint. *J. Theor. Biol.* 57, 355–371.

Sas, H. (coordination). 1989. *Lake restoration by reduction of nutrient loading. Expectations, experiences, extrapolations.* Academia Verl. Richarz, St. Augustin.

Scheffer, M. 1990. *Simple models as useful tools for ecologists.* Elsevier, Amsterdam.

Scheffer, M., Carpenter, S., Foley, J.A., Folke, C., and Walker, B. 2001. Catastrophic shifts in ecosystems. *Nature* 413, 591–596.

Schlesinger, W.H. 1997. *Biogeochemistry. An analysis of global change.* 2nd ed. Academic Press, San Diego.

Schramski, J.R., Kazanci, C., and Tollner, E.W. 2011. Network environ theory, simulation, and EcoNet(R) 2.0. *Environ. Model. Software* 26, 419–428.

Schrödinger, E. 1944. *What is life? The physical aspect of the living cell.* Cambridge University Press, Cambridge.

Shapiro, J. 1990. Biomanipulation. The next phase—making it stable. *Hydrobiologia* 200/210, 13–27.

Shelford, V.E. 1943. The relation of snowy owl migration to the abundance of the collared lemming. *Auk* 62, 592–594.

Shieh, J.H., and Fan, L.T. 1982. Estimation of energy (enthalpy) and energy (availability) contents in structurally complicated materials. *Energy Resources* 6, 1–46.

Shugart, H.H. 1998. *Terrestrial ecosystems in changing environments.* Cambridge University Press, Cambridge.

Simon, H.A. 1973. The organization of complex systems. In *Hierarchy theory—the challenge of complex systems*, ed. H.H. Pattee, 1–27. Braziller, New York.

Stevens, R.D., and Willig, M.R. 2002. Geographical ecology at the community level: perspectives on the diversity of new world bats. *Ecology* 83, 545–560.

Straskraba, M. 1979. Natural control mechanisms in models of aquatic ecosystems. *Ecol. Model.* 6, 305–322.

Straskraba, M., Jørgensen, S.E., and Patten, B.C. 1997. Ecosystem emerging. 2. Dissipation. *Ecol. Model.* 96, 221–284.

Svirezhev, Yu. M. 1990. Entropy as a measure of environmental degradation. Paper presented at Proceedings of the International Conference on Contaminated Soils, Karlsruhe, Germany.

Svirezhev, Yu. M. 1998. Thermodynamic orientors: how to use thermodynamic concepts in ecology? In *Eco targets, goal functions and orientors*, ed. F. Müller and M. Leupelt, 102–122. Springer, Berlin.

Sweinsdottir, S. 1997. Highest probability of life on the moon of Jupiter, Europa. *Ingeniøren*, no. 21, May, pp. 10–11.

Tiezzi, E. 2003. *The essence of time* and *The end of time*. WIT Press, Southampton, UK.

Tiezzi, E. 2006. *Steps towards an evolutionary physics*. WIT Press, Southampton, UK.

Tilman, D., and Downing, J.A. 1994. Biodiversity and stability in grasslands. *Nature* 367, 3633–3635.

Turner, D.B. 1970. *Workbook of Atm. Dispersion Estimates*. U.S. Public Health Service Publ. 999-AP-26

Ulanowicz, R.E. 1986. *Growth and development. Ecosystems phenomenology.* Springer-Verlag, New York.

Ulanowicz, R.E. 1997. *Ecology, the ascendent perspective.* Columbia University Press, New York.

Ulanowicz, R.E., and Abarca-Arenas, L.G. 1997. An informational synthesis of ecosystem structure and function. *Ecol. Model.* 95, 1–10.

Ulanowicz, R.E., Jørgensen, S.E., and Fath, B.D. 2006. Exergy, information and aggradation: an ecosystems reconciliation. *Ecol. Model.* 198, 520–524.

Van Donk, E., Gulati, R.D., and Grimm, M.P. 1989. Food web manipulation in lake Zwemlust: positive and negative effects during the first two years. *Hydrobiol. Bull.* 23, 19–35.

Viaroli, P., Azzoni, R., Martoli, M., Giordani, G., and Tajè, L. 2001. Evolution of the trophic conditions and dystrophic outbreaks in the Sacca di Goro lagoon (Northern Adriatic Sea). In *Structures and processes in the Mediterranean ecosystems*, ed. F.M. Faranda, L. Guglielmo, and G. Spezie, pp. 443–451, Springer-Verlag Italia, Milano.

Vollenweider, R.A. 1975. Input-output models with special references to the phosphorus loading concept in limnology. *Schweiz. Z. Hydrol.* 37, 53–84.

Wallace, A.R. 1858. On the tendency of species to form varieties; and on the perpetuation of varieties and species by natural means of selection. III. On the tendency of varieties to depart indefinitely from the original type. *J. Proc. Linnean Soc. Zool.* 3, 53–62.

Wallace, A.R. 1889. *Darwinism. An exposition of the theory of natural selection with some of its applications.* Macmillan and Co., London.

Warren, C.E. 1970. *Biology and water pollution control.* W.B. Saunders, Philadelphia.

Weiderholm, T. 1980. Use of benthos in lake monitoring. *J. Water Pollut. Control Fed.* 52, 537.

Wetzel, R.G. 1983. *Limnology* (2nd ed.). Saunders College Publishing, Philadelphia.

Willemsen, J. 1980. Fishery aspects of eutrophication. *Hydrobiol. Bull.* 14, 12–21.

Wolfram, S. 1984a. Computer software in science and mathematics. *Sci. Am.* 251, 140–151.

Wolfram, S. 1984b. Cellular automata as models of complexity. *Nature* 311, 419–424.

Woodwell, G.M., et al. 1967. DDT residues in an East Coast estuary: a case of biological concentration of a persistent insecticide. *Science* 156, 821–824.

Zaldívar, J.M., Cattaneo, E., Plus, M., Murray, C.N., Giordani, G., and Viaroli, P. 2003. Long-term simulation of main biogeochemical events in a coastal lagoon: Sacca Di Goro (Northern Adriatic Coast, Italy). *Cont. Shelf Res.* 23, 1847–1875.

Zavaleta, E. et al. 2010. Sustaining multiple ecosystem functions in grassland communities requires higher biodiversity. *Proceedings of the National Academy of Sciences.* 107, 1443-1446.

Zhang, J., Gurkan, Z., and Jørgensen, S.E. 2010. Application of eco-energy for assessment of ecosystem health and development of structurally dynamic models. *Ecol. Model.* 221, 693–702.

Zhang, J., Jørgensen, S.E., Tan C.O., and Beklioglu, M. 2003a. A structurally dynamic modeling—Lake Mogan, Turkey as a case study. *Ecol. Model.* 164, 103–120.

Zhang, J., Jørgensen, S.E., Tan C.O., and Beklioglu, M. 2003b. Hysteresis in vegetation shift—Lake Mogan prognoses. *Ecol. Model.* 164, 227–238.

Appendix

Atomic Composition of the Four Spheres

Element	Atoms % in (v.l. = very low)			
	Biosphere	Lithosphere	Hydrosphere	Atmosphere
H	49.8	2.92	66.4	v.l.
O	24.9	60.4	33	21
C	24.9	0.16	0.0014	0.04
N	0.27	v.l.	v.l.	78
Ca	0.073	1.88	0.006	v.l.
K	0.046	1.37	0.006	v.l.
Si	0.033	20.5	v.l.	v.l.
Mg	0.031	1.77	0.034	v.l.
P	0.030	0.08	v.l.	v.l.
S	0.017	0.04	0.017	v.l.
Al	0.016	6.2	v.l.	v.l.
Na	v.l.	2.49	0.28	v.l.
Fe	v.l.	1.90	v.l.	v.l.
Ti	v.l	0.27	v.l.	v.l.
Cl	v.l.	v.l.	0.33	v.l.
B	v.l.	v.l.	0.0002	v.l.
Ar	v.l.	v.l.	v.l.	0.93
Ne	v.l.	v.l.	v.l.	0.0018

Index

"f" indicates material in figures. "p" indicates material in problems. "t" indicates material in tables.